TOO FAR FROM HOME

TOO FAR FROM HOME

A STORY OF LIFE AND DEATH IN SPACE

CHRIS JONES

DOUBLEDAY • NEW YORK LONDON TORONTO SYDNEY AUCKLAND

PUBLISHED BY DOUBLEDAY

Published in the United States by Doubleday, an imprint of The
Doubleday Broadway Publishing Group, a division of Random
House, Inc., New York.
www.doubleday.com

DOUBLEDAY and the portrayal of an anchor with a dolphin are
registered trademarks of Random House, Inc.

All photographs courtesy of NASA.

LIBRARY OF CONGRESS CATALOGING-IN-PUBLICATION DATA
Jones, Chris (Chris Alexander)
    Too far from home : a story of life and death in space
/ Chris Jones.
        p. cm.
    1. Space vehicle accidents. 2. Manned Space flight—
Risk assessment. 3. International Space Station.
4. Columbia (Spacecraft)—Accidents. 5. Risk
management. I. Title.

TL867.J66 2006
629.45—dc 22

                                                    2006018748

ISBN 978-0-385-51465-1

PRINTED IN THE UNITED STATES OF AMERICA

1 3 5 7 9 10 8 6 4 2

FIRST EDITION

For Lee, who always brings me back.

CONTENTS

TOO FAR FROM HOME

PROLOGUE

Only minutes earlier, they had been something else—something big enough to be held in the hearts of millions—and soon they would be that big again, but now they were just three men in a bucket floating on the ocean, still far from home. Neil Armstrong, Buzz Aldrin, and Michael Collins had gone to the moon and back in the capsule nicknamed *Columbia* before splashing down 812 nautical miles southwest of Hawaii. Their miracle trip had taken them a little over eight days. It would be another three weeks before they'd complete the journey from the South Pacific into the arms of their wives.

In July 1969, the world changed, or at least its envelope did, pushed more than a quarter of a million miles across a vacuum. Even on a planet pockmarked by conflict, there was a new hope to latch on to. But that optimism didn't extend into every corner: no worry-minded scientist would gamble on how much these three men who'd changed the world had changed right along with it. Maybe they weren't like the rest of us anymore. Maybe they no longer belonged here.

They had lived in impossibly close quarters, drunk water from a pistol, and filled themselves up with a paste engineered to taste like Canadian bacon. They had been weightless, then not really, then weightless again, their blood still pumping but without the usual dams and anchors, flooding into their organs like water finding its level. They had crossed 25,000 miles in an hour. They had soaked up galactic radiation and navigated by stars. They had looked at snapshots of their families and swallowed hard, and they had wondered whether any single breath was meant to be their last. Two of them had walked in dust that might have contained spores, germs, bacteria, untold ancient lunar diseases and pandemics that every known inoculation couldn't fight; the third had passed over the dark side of the moon, out of radio contact, alone, for seven orbits, a hermit's passage.

Like no other men before, they had gone very far away. Who knew how different they might be when they came back?

Was something new and terrible hiding in the bottoms of their lungs or the ridges of their fingerprints? Or, worse, had they absorbed some stowaway parasite like sunlight through their skin?

What did space do to something as finite as a man's mind? How did punching a hole through Heaven unsettle a man's soul?

What kind of unforeseen reaction might begin if they dipped a foot into the salt of the ocean? If they shook hands with the rescuers who were on their way in the fat-bellied military choppers? Could even a sneeze make the 812-nautical-mile trip to Hawaii, and from there jump to Japan and California, choking billions of bronchial tubes with some nameless unstoppable plague?

How had space interrupted their bodies' clocks and rhythms?

How had it skipped their hearts?

How could it not?

And so for Armstrong, Aldrin, and Collins, the waiting began, first in their bucket, still far from home.

. . .

Back then, as forever, as always—until these days, perhaps—the remedy to any unexplored horizon was a colony. The men of *Apollo 11* would remain in their exclusive society, cut off from the rest of us, kept under glass. They would become the world's most famous and wide-smiling lepers. Three weeks seemed like a good settlement period. The mysteries of the universe would be waited out.

Every precaution would be taken till then. The swimmers dropped out of their helicopters and attached two orange life rafts to the module, one for decontamination and the other for recovery. One of the swimmers opened *Columbia*'s hatch, threw in three green, nylon, one-piece biological isolation suits, and slammed the hatch shut. Armstrong, Aldrin, and Collins each zipped on his suit. The American flag had been stitched to their left shoulders, their names across their chests; their faces were made alien by oval lenses and breathing masks. The swimmer then reopened the hatch and helped the astronauts into the decontamination raft. The four of them floated on the ocean's gentle surface, under clouds, looking in those outfits more like Martians than moonwalkers.

They were sprayed down with sodium hypochlorite (the module itself would get a betadine bath), transferred to the recovery raft, and lifted like tuna, in Billy Pugh rescue nets, into Helicopter 66 ("Old 66" to the Black Knights inside). The chopper normally hunted for lost surfers and enemy submarines off the California coast. Now its role had changed: for the astronauts in their zipped-up suits, it made more like a pretty good furnace. But the heat was a small complaint—during an earlier recovery exercise, high winds and seas had stalled the lift, and sharks had forced the swimmers back in their rafts. Now that there were no second chances, each part of the plan had to fit into the next without seams.

The decontamination raft and whatever invisible cargo it now harbored was scuttled, and the chopper made the short, thirteen-mile flight to the aircraft carrier USS *Hornet*, on which 2,115 officers and men, 107 NASA officials and civilians, a trio of pool reporters, and President Richard M. Nixon were waiting to make three men in a bucket big again. (The excitement had left the president first needing to take a leak. "Marine, where's the head?" was his opening verbal salvo after splashdown.) But no wives awaited. Old 66 touched down to cheers on the flight deck, was dropped by elevator to Hangar Deck No. 2, and Armstrong, Aldrin, and Collins walked through a plastic tunnel into their next new home, the Mobile Quarantine Facility.

Really, it was a souped-up Airstream trailer, a thirty-five-foot-long shining cylinder of unpainted aluminum, smooth except for the rivets. Inside, it looked like just about every other Airstream pulled off the assembly line, with the exception of an obtrusive ventilator above the fold-down table in the kitchenette. Isolation was guaranteed by negative internal pressure and the filtration of any effluent air. That was the science of it.

For the astronauts, though, it was just the latest in a long string of tin cans. Once inside, they showered, changed into blue flight suits, and settled in for speeches. Nixon, his bladder successfully emptied, told the three men that they had been the principal actors in "the greatest week in the history of the world since Creation."

Now joining the trio in their trailer were a technician named John Hirasaki and a NASA physician, Dr. William Carpentier.

The good doctor was never really part of the gang. There was a divide between him and the astronauts, the same gulf that's always broken off pilots from oddsmakers and logicians, flight surgeons especially. They had no dreaming in them. Asked what would happen if a medical emergency hit the crew of *Apollo 11* before they got off ship, Dr. Carpentier said, "That would be rough. But I'd say the Captain would have to treat the astronauts like carriers of an infectious disease and keep them in quarantine." The panic in a dying man's face viewed through a window would be trumped by the most pessimistic clinical imaginings.

With that grim scenario in mind, the crew of the USS *Hornet* began humping a souped-up Airstream to Pearl Harbor, full steam ahead.

. . .

In the meantime, the three travelers were subject to the first of several physical exams and asked to fill out customs forms, like any other tourists: in the space reserved for declarations, they wrote "moon rock and moon dust samples—manifests attached." Aside from border agents, thousands more islanders were waiting for them by the time they made it to Honolulu. As many as 25,000 hoped to catch sight of the fresh-tinned astronauts. The Mobile Quarantine Facility was lifted from the carrier, loaded onto a truck, and ferried through the waving crowds from the water to Hickam Air Force Base, where a U.S. Air Force C-141 jet transport waited to swallow the trailer whole.

They were back in flight, over the Pacific and on into Texas.

They touched down at Ellington Air Force Base in the early hours of the morning, with Armstrong providing the homecoming soundtrack on his ukulele. Finally their wives emerged out of the night. Jean Aldrin wore red; Pat Collins wore white; Jan Armstrong wore blue. They smiled up at their husbands, whose hands and faces were pressed against their palm-streaked windows to the world.

They were so close to home. But sometimes the distance between a man and his home can't be measured in miles. What keeps him away is time, or a wall as thin as a single sheet of glass.

. . .

The astronaut-lepers were hustled into the lunar receiving laboratory at the Manned Spacecraft Center outside Houston. It was as nice a prison as they could have asked for. They slept in genuine beds. Their showers were hot. They lined up for cafeteria-style food, and they ate together.

But they were prisoners, nonetheless. They were interrogated— they were asked what happened when, and sometimes they were asked the why of it, too—and they were poked and prodded, the way astronauts have been since they first touched space, shuffling through hallways with monitors strapped to their pale skin, eyed all the while by men hiding behind white masks as though something unearthly might burst out of their chests after any given breakfast. Fact was, no matter how much they tried to feel normal again, the rest of the world wouldn't let them. It began to dawn on them that their quarantine would never really come to an end.

Their feeling that they had become men apart went beyond all of the questions and examinations. It ran deeper than that. There were things that only they would ever know, things that they would never really be able to share.

They knew fear: there had been an even-money chance that *Eagle* would fail to lift off the moon's surface, leaving Armstrong and Aldrin to wait for their air to run out, as Collins watched helplessly, and that was just the start of the nightmare scenarios. They had told reporters before their trip that they had tried not to think about dying—in an explosion during the launch, or after colliding with a meteor, or by sinking in quicksand to the center of the moon, or because of something more mundane, like an oxygen leak, a guidance system failure, an uncontrolled spin, a fuel line plug, a cracked valve, or some goddamned shark waiting open-mouthed in the South Pacific—but they were also realists, and despite their brave public faces, they had gone through their wills before they left.

They knew, too, a terrible solitude: they had been planted in the middle of a desert in the middle of a blackness that stripped them of any horizon. They were as alone as men had ever been, cast in what Aldrin called "magnificent desolation," as if they had been sunk to the bottom of the ocean, with only the sound of their breathing for company.

Most unsettling of all, they knew longing, and for more than just their wives. From the moment they left the moon, it rose in them like a tide, minute by minute, day by day.

. . .

They thought it might subside, once they were back in their living rooms, once their long wait was over. Michael Collins, the trio's least-famous name but most-public face, summed up the feelings of the group: "I want out," he said.

At nine o'clock in the evening, on August 10, 1969, they got their wish. At last they hugged their wives, smiled for the cameras, and headed home. They sat on their couches, and they put up their feet. A record amount of tickertape would soon fall on them in New York City, the big-blast kickoff to the rest of their now-historic existences, lived out in the world's memory banks and on free luxury cruises, in exchange for giving small talks. It would all be very fine.

But it would never again be enough. Worst fears had come true. They really weren't like the rest of us anymore. Space had changed them after all, only in ways that science might not have predicted and Armstrong, Aldrin, and Collins might never have dreamed.

For the rest of their lives, no matter how many crowds surrounded them or how much tickertape fell on their heads—no matter how many hearts they were held in—in their own hearts they would remain three men sitting in a bucket, forever far from home.

1 SIMPLE MACHINES

For this one dream, men had turned chimpanzees into crash test dummies, gone through a thousand pink enema bags to make sure their own plumbing was ready to withstand the trip, and finally been launched like artillery shells—in corrugated-tin capsules held together by hardware-store screws—deep into the black. Not much later, they were balancing themselves on top of six million pounds of rocket fuel and lighting it on fire. Today the insanity physics continue. Astronauts blink down the risk that a rubber O-ring on one of the space shuttle's solid rocket boosters might give way, spraying a flame laced with powdered aluminum, ammonium perchlorate, and iron oxide onto the external fuel tank, igniting its cargo of liquid oxygen and hydrogen, and having their cockpit turn into a coffin.

All to cross the gap between home and away, to cross a distance that, on land, any old rust bucket could fart across in a couple of hours. But the gulf between earth and space is, and always will remain, a wider divide: it's a chasm without walls, and plenty of men, as well as a couple of women, have died trying to string their way to the other side.

. . .

Captain Kenneth Bowersox had survived the trip four times, twice as a pilot in the space shuttle's forward right seat, twice as commander in the forward left. Now he played the unaccustomed role of cargo, staring at rows of storage lockers instead of the beckoning sky. The pilot had become the passenger, one of three men crammed below decks like ballast, waiting to be shuttled on *Endeavour* to the International Space Station.

Despite having been shunted downstairs for launch, Bowersox had been looking forward to his fourteen-week-long mission the way the rest of us look forward to a much-needed vacation. Although he had visited space four times, none of his previous shuttle missions had lasted more than sixteen days, and he had never been to the International Space Station. He had always felt that he had been asked to come home too soon. This time, however, he would have time to linger. He and his colleagues would conduct a range of scientific experiments and busily maintain station—astronauts rarely bother to slip *the* in front of *station*, thinking of it as a place rather than a thing—but their principal assignment would be to make themselves and the men and women who would follow them content living in orbit. Even before launch, Bowersox was confident that, as far as finding happiness went, he would succeed. He might have been flying steerage, but space was still his island in the sun.

For all that Bowersox tried to focus on the destination, he couldn't help wishing he was up above for the journey. He wished he was alongside the two men in the front-row seats—in *his* seats—able to take in the view and, more important, see the fifty control panels and nine monitors that flashed before Commander Jim Wetherbee and Paul Lockhart, the pilot. Against his life's habit, Bowersox had ceded control, and now he shifted in his seat and fiddled with his straps. At least Wetherbee had been in space five times already, and like Bowersox, he was a Naval Academy man and okay by him; Lockhart, in contrast, was making just his second trip, and only five months after his first, back in June 2002.

Also, he came out of the air force.

Worse, Lockhart wasn't meant to be flying today. Had everything gone to plan, Lockhart should have been in Houston, watching NASA TV, trying to get out from under the private jealousy that runs through every grounded astronaut forced to watch another man's dreams come true.

The man stuck watching television this time around was Gus

Loria of the marines, who had thrown out his back in August and been scratched from the mission, which would have been his first. Instead, Lockhart's vacation plans had been canceled, and he was pressed into emergency service, jammed into the same seat on the same shuttle he'd occupied just that past summer. It was still set for his height, and he settled right in.

Loria was less comfortable on his perch back in Houston, and he wasn't alone among the unhappy spectators. Joining him was Dr. Don Thomas, a four-trip veteran and the science officer who had been expected to join Bowersox and the Russian cosmonaut Nikolai Budarin—a former engineer who had logged nearly a year in space on Mir, the International Space Station's burned-up predecessor—for their stay. Over two years of training, at home and in Russia, in simulators and classrooms and T-38 jets, they had become Expedition Six.

Thomas had also undergone a more sinister indoctrination. Without the apron of earth's atmosphere to protect them, astronauts are exposed to higher-than-usual amounts of solar radiation. Because little is known about exactly how much exposure will trigger cancer, and rather than risk its astronaut corps becoming lumpy with tumors, NASA has set an arbitrary radiation "red line." If an astronaut approaches that ceiling, he's grounded and stuck behind a desk until his cancer-free retirement (fingers crossed). Extensive medical investigation had revealed that Thomas, for whatever reason, had come unacceptably close to NASA's red line. Another four months in space and he would have gone over it. He would have carried too much of the universe home with him.

The flight surgeons had passed on their findings to Mission Control and, in turn, to Bowersox. As the commander of Expedition Six, he had been left facing down three possible outcomes following the unsettling news: he could choose to ignore the evidence and fight to allow Thomas to fly; he could see Thomas scratched from the mission and replaced with his designated backup, a chemical-engineer-turned-rookie-astronaut named Don Pettit; or Bowersox could ground himself, Budarin, and Thomas, and order all three members of Expedition Six replaced by their reserves. He had taken the options to bed with him and been surprised by how much time he spent turning them over.

Through training and by nature, Bowersox had acquired a certain cool. He carried a sense of detachment with him almost always: a pilot's life, if he wants to see the end of it, doesn't hold a lot of room for romance, and Bowersox had mastered the hard art of bottling up his feelings. Confronted with a dilemma that would keep most men up at night, he'd hold it under the light like a clinician, pulling it apart without emotion. The walls he'd built carried clean through his eyes, which were the same hard, glacier blue that had become a trademark of the best pilots, like Chuck Yeager's drawl or a strong chin. (Bowersox, who grew up in Indiana, owned the chin but not the accent.) Since Norman Mailer had pointed out that all but one of Apollo's first class of sixteen astronauts boasted blue peepers, that genetic fluke had become a virtual requirement of the astronaut corps. It was as if the color of a man's eyes revealed the tenor of his heart, cold and colder.

But here Bowersox struggled, even though the facts were plain. Thomas's health presented a risk, and a trip into space was marbled with enough risk already. That should have been all there was to it. And yet, for one of the few times in his life, it was finally his turn to lie awake, allowing the data to be clouded by late-night sentiment. He had grown to like Thomas—a quiet, hardworking, serious-minded man, the sort whose hands never shook. Bowersox's affection for him, when viewed through the peculiar prism of space travel, was a particular kind of love: it meant that he was both comfortable in his company and confident in his abilities. They had developed an abiding faith in each other, and now Bowersox was confronted with a decision that, in an instant, might break what had taken years to build.

He didn't want his friend killed with kindness, however, and he began casting his mind toward switching out the entire crew. It didn't take him long to shake off that option like a shiver. The clean sweep would have crushed Budarin and brought Thomas no closer to space. And in the honesty of his private company, Bowersox had to admit that his own itching to fly bordered on a sickness. Through the semidarkness, he stared down the prospect of spiking what might be his last stab at it. He was forty-five years old, almost forty-six, growing long-toothed by astronaut standards; he'd lost his ginger hair a long time ago. Deep down, he knew his time was running out. He also knew there were dozens of astronauts lurking in the wings behind him, first-stringers their entire lives who'd found themselves in the unnatural position of waiting, sometimes for seven, eight, nine years, hoping that their phone would finally ring with the call that gave them the go-ahead. No part of Bowersox wanted to put a line through his own name in exchange for one of theirs; no blue-eyed pilot would ever volunteer to give up the stick.

All of which had left him with a single option: replacing Thomas with Pettit, exchanging one Don for another, and, in the process, learning how to think of a friend as though he was just another part of the machine.

. . .

At Star City, an hour north of downtown Moscow, down a road cut through a green forest, a contingent of exiled Americans had gathered in the small cottage occupied by Don Pettit. He had been in Russia for more than a year, mostly going through the motions. Although he took his training seriously, he knew that, as a reserve, his chances of getting called up to join Expedition Six were close to zero. Really, his agreeing to a semipermanent exile was part of a grander plan he had drawn up for himself. For a rookie astronaut, clocking in as a backup was viewed favorably by those few, untouchable men in Houston who put together crews. So long as Pettit performed well enough in training, and providing he didn't do anything that might make the Russians wary of him, he would earn a better than average chance of one day making the trip to station. Until then, he would uncomplainingly do his chores, biding his time as though serving a prison sentence, pushed along by the hope that perhaps Expeditions Nine or Ten or Eleven might include him, front and center.

Pettit looked the part, at least, every inch the science guy—

glasses hiding brown eyes (not blue), curly dark hair, an affinity for cargo pants held up by a belt full of tools. He was a chemical engineer, an inventor, a veteran explorer of molecules and optics rather than of space, a man who couldn't help wondering how engines worked, why clouds formed, what lived in the hearts of volcanoes. In his endless quest to understand more about the inner workings of the universe, he had tried and failed to become an astronaut three times; the fourth time around, he was finally given the chance to dissect the stars.

To fill the hours until he made the jump from reserve to prime, he hosted loud parties in his cottage, especially when his wife, Micki, and their tiny twin boys made the flight over for a spell. She was a singer, and along with some of Pettit's astronaut colleagues—including Chris Hadfield, the amiable Canadian guitarist—had formed a band. Late one night in August 2002, they had taken seats wherever they could find them, on the floor and the couch, and they had played and sung and laughed until they were interrupted by the phone ringing, not long before midnight. The noise in the room stopped. Pettit answered, and after he had listened to the calm but serious voice on the other end of the line, he hung up the phone, shot Micki a look, and rushed out the door.

He had been told a few days earlier that there were "anomalies" in Don Thomas's medical evaluation, but nothing more specific. The news had been passed along as a courtesy more than anything else. Hiccups were not unusual, and Pettit had never thought, at least not for more than a moment, that this minor tremor might become an earthquake. But by the time he had returned to his cottage—by the time Micki had the chance to lay her eyes on him again—she knew what he knew: in three months, both of their hearts would thump through their chests, counting down the seconds to liftoff and a long time away.

· · ·

Ken Bowersox's decision was not clean in its consequences; one dilemma begot a dozen others. First, Thomas's clothes and food had been shipped ahead to station. His set of embroidered blue golf shirts had the right first name stitched on their pockets, but the taller Pettit would need to pack along his own pants and sneakers. More troublesome from Pettit's perspective, Thomas—like Bowersox and Budarin—had forgone coffee in his food allowance, a hand-picked menu served on an eight-day cycle. Pettit, who liked to kick-start his day with a jolt of caffeine, begged for permission to carry up some coffee. After threatening tears, he was allotted about one hundred bags of freeze-dried instant; because the cost of shipping cargo into space runs about $10,000 a pound, he was lucky to get that much. (A fan of spicy food, Pettit was also permitted a dozen cans of New Mexican green chiles to dress up Thomas's humdrum choices.)

Pettit's more immediate concern was Thomas's emotional health. His grounding had left him gutted. Thomas had fought the findings as soon as they were announced; the scientist in him had always loathed the "red line" that ultimately did him in, railing against it as so much hokum theory. He believed in *evidence*, in hard arithmetic and indisputable sums, and now, in his mind, all of the time and hope that he had invested in this mission had been wiped away by calculations fraught with doubt. In the weeks that followed, after he had returned to Houston and sat alone with the lights out, his mood had continued to swing from anger to upset, the spaces in between occupied by a kind of disbelief, those sad moments when he tried to convince himself that he could change his fate and win his return to space.

Switch-outs for still-living crew are rare, much rarer than replacing the recently deceased—a grim reminder that pushing the limits of astronautics is usually an all-or-nothing proposition. Their scarcity had made them the ultimate bad omen, even in a profession routinely beset by metaphorical broken mirrors and black cats. Over the course of space travel's voodoo history, the next man in line had replaced Elliot See, Charles Bassett, David Griggs, and Sonny Carter after each had been killed in an air crash before his scheduled launch. But before Thomas and Loria had lost their spots, bad news had been delivered to an astronaut rather than to his wife only twice. Deke Slayton's irregular heartbeat bumped him from

Mercury's flight order in 1962. And more famously—thanks to the blockbuster film—Tom Mattingly was replaced by Jack Swigert after he had been exposed to the measles before the ill-fated flight of *Apollo 13* in 1970. Bowersox had seen flashes of the movie in his head when he had dropped the bomb on Thomas. He marveled at how much harder real life played out than it did on film, all the while trying not to fixate on the fate of the last crew broken up so close to launch.

Swigert had joined Jim Lovell and Fred Haise, and they had been none too happy for his company. Unfortunately, he also happened to be the man who flicked the switch to stir the oxygen tanks in *Apollo 13*'s service module on its way to the moon. Because of an earlier, long-forgotten mishandling of the No. 2 tank—it had been dropped and replaced during *Apollo 10*'s kitting out—exposed electrical wires shorted and lit the tank's Teflon insulation on fire. The oxygen was slow-boiled, the fire spread along the wires to an electrical conduit, and the tank blew up. The explosion damaged another oxygen tank and the inside of the service module, and it ejected the bay No. 4 cover into space: in terrible sum, it put a hole in the machine. Although the crew of *Apollo 13* somehow managed to limp their way home on courage, they were destined to become part of astronaut lore for different, darker reasons. Their preflight drama, coupled with their mission number, meant that their lessons were the kind passed on in whispers. When it came to catapulting yourself into space, there was no such thing as superstition. There were only signs.

· · ·

For Expedition Six, the signs continued to suggest that they might be better off staying home. On October 7, their sister shuttle *Atlantis* had a close call when a set of explosives—designed to blow apart the eight giant bolts that pin down the vessel until launch—failed to detonate. *Atlantis* still lifted off because another set of explosives had done its job, but the misfire raised alarms and caused onboard computers to seize up, forcing controllers on the ground to override automatic systems. More worrisome, no one could figure out in the aftermath why the charges hadn't tripped. Workers went to the trouble of replacing wiring harnesses and electrical connectors on the launchpad, but in a lot of ways, that work was helpless. It was a blind stab at solving an unknown problem. When it came time to let loose *Endeavour*, no one could guarantee that the right kind of blast was about to take place.

A little more than a week later on the other side of the world, the wrong kind happened. At the Plesetsk Cosmodrome in northern Russia, an unmanned *Soyuz-U* booster became a fireball about twenty seconds after liftoff, killing a soldier on the ground and injuring eight others. An investigation found that metal contamination in the rocket's hydrogen peroxide system had triggered the disaster. Russian officials wondered openly whether the fatal flaw had been the work of terrorists. At the least, the accident delayed the launch of a *Soyuz* taxi mission to the International Space Station, which pushed off the ferrying of Expedition Six from the early morning of November 10, 2002, until shortly after midnight on November 11.

Then and there, cast in spotlights, *Endeavour* would be waiting for them, looking from a distance like a monument to miracles and up close like a bottle rocket.

Always, it had been a little bit of both.

. . .

The space shuttle is a complex jumble of bones and arteries, but at its heart, it's a gas tank. The majority of its juice is bottled up inside the massive rust-colored external tank strapped to the shuttle's underbelly. At 153.8 feet long and with a diameter of 27.5 feet—the size of a Boeing 747, the plane that the shuttle sometimes hitches a ride on—the tank dwarfs the vehicle it fuels. A car's gas tank is about 5 percent of its total mass; a fighter jet's is about 30. The shuttle, including the two solid-rocket boosters locked to its sides, is 85 percent propellant. It's 1,107,000 pounds of powdered aluminum mixed with oxygen off-gassed by ammonium perchlorate, and, in the external tank, it's another 143,060 gallons of liquid oxygen and 383,066 gallons of liquid hydrogen, good for an additional

1,585,379 pounds of spark. Upon ignition, they combine in dual pre-burners to produce high-pressure gas that drives turbopumps in the shuttle's three engines. The rest of it is torched in the main combustion chamber, which reaches a temperature of 6,000 degrees Fahrenheit.

In the anxious hours before that last button is pushed, the hydrogen and oxygen will have been supercooled and pumped, very carefully, into the tank. Exhaust vents work at preventing rupture, but even with every precaution and an inch of insulating foam at work, the tank's aluminum housing creaks and groans under the pressure, sounding like an icebound lake breaking up in spring, like whale music.

. . .

On November 10, at 9:35 p.m., *Endeavour*'s seven-man crew answered that call.

Wetherbee and Lockhart had readied themselves to fly. For Wetherbee, the first American to command five space missions—by the end of this voyage, he was scheduled to have logged more than 1,500 hours in space—the preparation for launch felt close to routine, or as close to routine as rocketing into space ever could. It helped that he had completed a nearly identical mission in March 2001, having ferried Expedition Two to the International Space Station and brought Expedition One back to earth.

Despite Lockhart's late substitution, he had also found comfort in his unexpected mission, and not just because of his custom-fit chair: his single previous shuttle flight had exchanged Expeditions Four and Five.

In addition to helping Wetherbee guide the shuttle toward station, Lockhart was charged with coordinating the space walks planned for the two men seated immediately behind him. Mission specialists Mike Lopez-Alegria (the third Naval Academy graduate on board) and John Herrington (the first tribally registered Native American tapped to fly into space) would need to head outside three times after docking, continuing the construction of the still-expanding station. Along with his tools, Herrington carried with him six eagle feathers, a braid of sweet grass, two arrowheads, and the Chickasaw Nation's flag. A native of Madrid, Spain, Lopez-Alegria—"Mike LA" to his crewmates—also had the hopes of an entire people resting on his shoulders. Like Wetherbee and Lockhart, he had visited station once before, becoming something like a celebrity after his appearance in *Space Station 3D*, an IMAX documentary narrated by Tom Cruise.

Meanwhile, Expedition Six—Bowersox, Budarin, and Pettit—had finished resigning themselves to disappearing mid-deck, hiding out like stowaways, like kids sneaking into a drive-in by getting locked in the trunk of a car.

Assigned seating aside, the seven men remained equals in the most important respects. All of them shared the burden of foreboding during the traditional prelaunch supper that had been prepared for them, one last meal off plates. The heroes of *Mercury* and *Gemini* and *Apollo* would tuck into something suitably stout, steak with liquid centers and eggs over hard, but on that day, not everyone had an appetite. The less that went in, the less that could come out, and no one wanted to be the first to throw up.

Next they returned to their private quarters. Outside of their rooms, a flight diaper and what looked like long underwear—a full-body garment strung with hoses that would be filled with cold water to wash away the heat of the moment—were waiting for them, and they put them on. Then they each walked to a room lined with burgundy recliners.

There they were helped into eighty-six pounds of spacesuit, not including their helmets and gloves. All of it was designed to improve their chances of survival, with or without an accident, and it was hard to escape the feeling that they were dressing for danger. Their armor and shields included an integrated exposure suit, a parachute harness, and flotation devices; the big pockets on their legs were stuffed with survival gear; even the bright orange color of the suits was a nod to safety, because it would make the astronauts (or their bodies) easier to spot if they were ditched into the ocean.

The spacesuits were relatively new inventions, changed up and bolstered after *Challenger* had come apart seventy-three seconds into its flight on January 28, 1986—and after Joseph Kerwin, a former Skylab astronaut and a biomedical specialist in Houston, determined that the crew might have survived the initial explosion. "The forces to which the crew were exposed during Orbiter breakup were probably not sufficient to cause death or serious injury," he wrote in his final report. He did leave the hopeful opening that the crew might have been unconscious had the cabin lost pressure, "but not certainly." He regretted to note that there were several troubling signs that they were, in fact, aware of their fate, including the activation of three personal egress air packs connected to the crew's helmets, which had to have been turned on manually. That raised the specter of the seven lost astronauts having been very much alive during their freefall into the Atlantic Ocean, killed only by the impact of splashdown.

Every astronaut who has followed their footsteps to the launchpad has imagined the two minutes and forty-five seconds it took for them to hit the water. Every astronaut has taken the time to wonder how they would have filled it.

To help *Endeavour*'s seven-man crew overlook the horror of their imaginations, they had the option of tucking away a good-luck charm or a talisman, but few of them did. (Most of them remembered that after the hatch had popped open on his floating capsule, *Mercury* astronaut Gus Grissom had nearly been pulled to the bottom of the ocean by his pockets full of souvenir dimes.) Instead they relied on deep breathing to get them through. Bowersox and Budarin, both experienced fliers, were pictures of calm. After he was dressed, Pettit tried to emulate them, leaning back in his chair and putting his hands behind his head, closing his eyes, exhaling slowly. He tried to look as though he was at home, bunking down for a nap on the couch, but there were wake-up calls everywhere he looked. Closest, a plastic band snapped around his wrist reminded him of his blood type and what medications he was allergic to. He hoped that no one else would need to know.

After everybody was suited up, together they walked down a long hallway lined with technicians and staff shouting encouragement, one last charge of adrenaline that felt as though it might have pushed them into orbit all on its own. They took an elevator down to the ground, passed through a set of metal doors, waved again to the assembled press, flashbulbs popping, and climbed into a silver bus called the Astrovan.

It was a ten-minute drive to the launchpad, almost two minutes longer than their ride into space would take. Their path was cleared by a security helicopter and armed escorts. A few of them took the time to whisper the astronaut's prayer under the din: "God help you if you screw up."

About a mile from the stack, with the shuttle looking huge and beautiful, lit up and calling out to them again, a security guard waved his flashlight through the gloaming, stopping the bus. He climbed on and asked the astronauts to show their security passes. Six of them pulled out laminated cards with their mug shots and authorization. Don Pettit, so close to his dream, patted himself down for a pass that wasn't there. He hadn't even seen one before, he thought, and besides, zipped up in his spacesuit, he was clearly supposed to be here. If his costume was a counterfeit, it was a perfect one. But with the guard working his way down the aisle, Pettit's voice was close to cracking when he began to apologize, stammering that he'd somehow missed this step along the way. The rest of the crew turned back and stared him down, eyes rolling. After the bumps on the road to their launch, this, it felt like, was the most calamitous. The shuttle was close enough for them to run to, and here Pettit, the rookie reserve, was going to be pulled off the bus because he didn't have this cheap piece of plastic to show some puffed-up rent-a-cop. He was stared at just long enough for him to grow frantic before Wetherbee finally cracked a smile, and then Bowersox did, and soon enough the whole bus was broken up in laughter.

Even at a time like that—perhaps especially at a time like that—there was room for a joke. Pettit, dying inside, tried to squeeze into the space under his seat until they arrived at the pad.

Once herded off the bus, they took the elevator 195 feet up the shuttle's hull, watching all the while condensation running down the sides of the external tank, falling into the trench that would catch their fire beneath them. Finally they found themselves in the White

Room, a closed-off sanctuary in which they finished the last of their waiting. (Three miles distant, their families would find them by the bright light.) One by one, helped by six technicians in white suits and ball caps, they crawled through the hatch on their hands and knees.

Ten minutes later, they were on their way back to the ground.

Earlier in the day, a valve had been opened, allowing the oxygen that would be pumped into the crew's helmets and cabin to flow through the ship. Now routine preflight tests had found a small amount of that oxygen in the shuttle's cargo bay. There was no good reason for it to be there. Somewhere in the bundles of flexible hoses under the floor, there was a leak.

"Tonight's not our night," Steve Altemus, NASA's launch manager, had crackled over the radio. "I know you guys are going to be disappointed, but I think we want to give you a healthy vehicle before we cut you loose from the Cape."

"Absolutely," Wetherbee said.

And that, for the moment, was the only absolute. As with the faulty bolt explosives, no one was sure exactly what the problem was. No one yet knew how to fix it. Here was this giant, groaning stack of metal and ceramic tiles and rocket fuel, and some virtually invisible thing in it had gone wrong. It was probably something painfully small, no bigger than a pinhole. But in space—in a vacuum without gravity—small things grow into big things, and a pinhole is plenty big enough to leave seven men trapped in a box without air.

. . .

Thousands of hands guide the shuttle on its journey to liftoff. Like the fibers of a wire, more than one hundred private aerospace contractors and subcontractors across the country conspire to muster the necessary current. The Boeing Company of Chicago and the Lockheed Martin Corporation of Bethesda are the principal circuits; through their jointly owned subsidiary, United Space Alliance of Houston, they've been responsible for the day-to-day operations of the shuttle since 1996. Among their suppliers are ATK Thiokol Propulsion of Brigham City, Utah, which builds the solid rocket boosters; Spacehab, Inc., of Webster, Texas, which designs and manufactures the modules that house experiments; and United Technologies of Hartford, which, through its Pratt & Whitney engine divisions in Florida and California, forges the turbopumps. The external tank is welded in Michoud, Louisiana. The life support system comes out of Windsor Locks, Connecticut. The main engines were first brought to life in Canoga Park, California.

At each of those sites, dozens of processes combine to produce a single part of the shuttle or sometimes only a part of a part. For instance: the shuttle's nose cap can withstand temperatures as high as 3,000 degrees Fahrenheit, but it doesn't provide much insulation for the crew looking out over it, and in its hollow core, a bundle of thirty-two heat-resistant blankets must be packed into place. Every one of them is made from scratch at NASA's Thermal Protection System Facility. Ceramic fabric is first measured and cut and coated with sizing to prevent the fibers from coming apart; the fabric is layered between insulating batting and stitched; it's trimmed and sewn closed around its edges; the completed blanket is finally baked twice in superhot ovens and waterproofed. The process takes a team of workers two months to complete—all to produce a single critical thing in a machine born from thousands.

By barge and plane, truck and train, the pieces are shipped to the Cape and made whole in the Vehicle Assembly Building, a leftover from the days of Apollo. Its volume is almost twice as large as the Pentagon's. To put that sort of scale in perspective, 6,000 gallons of paint were needed just to tattoo an American flag on its flank. In it, a second army of workers picks up from the first, taking the parts and making them into a whole, a machine of almost surprising fragility.

They all remember that for STS-71—the one hundredth manned flight in American space history—a small number of Northern Flicker woodpeckers had taken roost on the shuttle *Atlantis*, knocking holes into its external tank. Since then, fake owls have been installed around the launchpad, leaving this fantastic, $2 billion spaceship guarded by a few bucks' worth of Taiwanese plastic. Those owls serve as a constant reminder of how close failure is, how everything and everyone here depends on everything and everyone else, the bunch of them tied together like climbers roping their way up a mountain.

And if any one of them misses a step, or if any of the thousands of others before them already has—if an O-ring gets too cold and brittle, or a detonator cable is left unattached, or a sliver of aluminum finds its way into the hydrogen peroxide system, or an oxygen tank is dropped and forgotten about and then a switch is flicked on the way into space—the shuttle and seven astronauts will be lost, probably in a ball of fire and smoke.

. . .

In the hours after the oxygen leak was discovered, workers emptied the fuel tanks as carefully as they had filled them and opened the payload doors, hoping to find the source of the oxygen leak that had sent the crew back to the ground. Bag-suited engineers climbed aboard a platform that would lift them to the front of the cargo bay to start their inspections. Just as they began to rise, a spotter on the ground was distracted, and the platform bumped into the Canadarm, the shuttle's robotic arm. A small square of the arm's thermal protective blanket was torn away, and now there were two problems to fix, not one.

If the arm was ruined, the *Endeavour* would need to be rolled off the launchpad, pried loose from its external tank, towed back to its hangar, have its arm replaced, and finally get fitted out for flight again. That kind of delay would likely have pushed liftoff into something called the beta-angle period, a two-week stretch in December when the sun and the earth conspire to leave Expedition Six's ultimate destination, the International Space Station, without shade. While the station can rotate and shift position to protect itself from the heat, the shuttle can't stay docked in the middle of those acrobatics. *Endeavour* would remain grounded until the end of the year.

For the crew, it was one more worrisome hitch. In their private quarters, still coming down from their near-launch adrenaline burst, they were called together and told that their next try would be delayed until November 18 at the earliest, and that they might as well head back to Houston. They packed up their few belongings and flew home in their trusty two-seater T-38s, feeling disappointment and maybe just a little relief. They had been granted a reprieve from deadline's stress, if only for a short while.

They remained largely locked down in the quarantine that they had been ordered into weeks before. Contrary to the feelings of the *Apollo* astronauts, the enforced isolation was not just for appearances. One common cold shared among them might have been enough to ruin everything. (Already, sinus congestion is the plague of most missions, because fluids that are normally drawn down by gravity suddenly start flowing up.) But even for the toughest-nut crew members, the exile was hard to stomach. These were boring, idle hours filled with last-minute busywork and the occasional outbreak of night sweats. It was as though the men had been given their destinies in gift-wrapped packages and then were told that they had to wait to open them . . . and wait . . . and wait . . . and wait.

For their patience, they were each rewarded with eight hours in their private bungalows and split-levels immediately after they returned to Houston. Pettit sneaked home and saw Micki, but he made the visit late at night, when he was sure his boys were asleep. He opened their bedroom door. Their faces were barely illuminated by the light in the hall. He stole a look at them, but he didn't dare even whisper a good night or a farewell, lest he wake them up and hear their cries for a hug. Then the father and the astronaut in him would have had to fight it out, and he didn't like to think about who might have won.

. . .

Back in Florida, there had been no such break. A flurry of work had begun, aimed at beating the beta-angle period and putting an end to Houston's waiting. Using ultrasound equipment, sleuths found a bruise on the carbon composite material that makes up the Canadarm's bone. Engineers in Toronto replicated the damage on a working test arm and began running a series of experiments, trying to decide how the wound affected the arm's structural integrity.

Meanwhile, workers at the Cape found the source of the oxygen leak: a small metal part—another one of those single critical things—had worn out and cracked a hose near the cargo bay's forward bulkhead, just behind the crew cabin. It was replaced and the rest of the hoses were tested.

There was more good news. The Canadian engineers decided that their arm was in fine working order and needed only a patch.

With that work under way, NASA decided to open a launch window. The moonlit evening of November 22 looked like a good, safe bet.

. . .

The crew spent their last, long days before launch filling the lonely hours the way astronauts always have. They had trained as though for inevitable disaster, strapped into simulators that replicated just about every physical possibility: the motion-based simulator, which swung them through every axis to prepare for the turbulence of launch; a fixed-base simulator that left them able to sketch from memory the shuttle's flight deck on the back of a napkin; an engineering simulator to practice the fine art of rendezvous; Shuttle Training Aircraft to make rote even the most difficult approach; T-38 flights to keep up on their instrument reading; and the Neutral Buoyancy Laboratory for space-walk training and emergency bail-out practice. They'd even exited a shuttle mock-up in wire baskets and learned how to drive the tank that sits on the launchpad, its armor thick enough (theoretically) to protect the crew in the event of catastrophe, providing they made their way to it in time. But now that make-believe was giving way to reality, all of the training in the world would fail to carry them into space. Now that they weren't just playing pretend, it was time to plumb some deeper well.

All three members of Expedition Six were married. Bowersox had three children, the youngest of whom was six; Pettit's twins were not yet two years old; Budarin had a pair of teenage sons. After making sure their wills were written and in place, they penned letters and cards for their kids that they hoped would never be read, outpourings of sentiment that, depending on the course of the coming days, might become the only memory that sons had of fathers. Imagining those letters being carried in the wallets of boys grown into men, or folded up in their nightstand drawers, or hidden inside boxes of secret things . . . That was enough to make even the most sure-minded man feel a lot less like saying goodbye.

Not entirely by accident, leaving was no longer a choice. Once again, they flew across the Gulf from Houston to the Kennedy Space Center. Their families were waiting for them there, but behind a double yellow line that no one was permitted to cross in case germs came with them. There would be no more hugs, only smiles and waves. Sometimes there isn't even a single pane of glass between an astronaut and home. Sometimes only an idea and some paint get in the way.

They boarded a bus that took them from the landing facility to the crew quarters to catch some rest, maybe even a little sleep. And then they went through the entire routine again, starting with an uneasy dinner and ending with a tight squeeze through the hatch.

Inside the crew cabin, one last hand-up was waiting for them, from an astronaut-turned-technician called the Caped Crusader. He helped each man into his seat. For each of them, one by one, four parachute clasps were done up. Four seat clasps were buckled. Oxygen hoses were attached, helmets were put on, communications lines were plugged in, and headsets fired to life. In the scramble, Pettit's bag of coffee was taped under his seat, and Budarin managed to spruce things up a bit, tying a windup toy bee on a string to one of the locker doors in front of him. (It was strange for Pettit to see a barrel-chested man who looked like Charles Bronson, with his deeply lined forehead and head of thick hair, fumbling with a child's plaything just then.) At last, the nervous work was done. Everybody was tucked away more than two hours before launch. Cold water began running over their bodies through their wired-in undergarments. It felt like jumping into a swimming pool on a hot day.

They were given one last wish of good luck. The hatch was closed. And that was it. That was the last of the lingering.

Except. Across the Atlantic Ocean, at two small towns in Spain, the weather was bad. In Morón and Zaragoza, clouds rolled in and the winds picked up. It wasn't an inconvenience solely for holiday-makers. Both towns are home to air force bases; both of those have been designated Transoceanic Abort Landing Sites for the shuttle. If one of the three main engines fails, or if some other critical component breaks down and makes entry into orbit impossible, the shuttle's commander has between eight and fifteen minutes to try to return to the launch site, to slingshot his way around the globe and touch down in California, or to set his sights on the middle ground of Spain, where about sixty pilots, astronauts, technicians, and medical emergency personnel had gathered alongside an otherwise empty runway. Just in case, they checked their watches and fired up ambulances and sent a weather balloon into the sky. On this early morning, however, they didn't need the balloon to give them the forecast. Just as the launch window opened at the Cape, the teams at Morón and Zaragoza watched lightning flash across black skies and shook their heads.

The countdown stopped. Bowersox, Budarin, Pettit, and the rest of *Endeavour*'s crew were helped out of the cockpit.

Fate would have to wait once again. Better luck tomorrow.

. . .

After all of which—after all of the switch outs and misfires and delays—even the cold-eyed Ken Bowersox spent the following evening trying to keep his heart rate within acceptable levels. It hadn't helped one bit, he muttered to himself, that their mission number was STS-113. The crew had whispered to one another about asking for a new designation, perhaps moving on up to STS-114, jumping the way hotel elevators skip that cursed floor. Maybe it was all that they needed to end their bad run, like a baseball player changing his socks to break out of a slump. But in the end, pride subsumed the talk of jinxes; they decided to swallow their ill feelings. If outsiders somehow caught wind of their conversations, they agreed that they would brush them off as a joke, a defense mechanism, a distraction to lighten a somber mood. And yet deep down, there remained an unease in them, a low, unshakable hum in the background. It wasn't fear, and it wasn't despair, and it wasn't resignation. It was a creeping anxiety, a kind of shadow. Having been suited up for the third time, a few of them wondered whether they might never leave the ground. They wondered whether this whole big, crazy thing wasn't meant to come off. Now Bowersox stared at the lockers in front of him and tried to push aside the last of his own wonder, about whether they should have changed their mission number after all.

"Looks like we've got a good vehicle and good weather tonight for you," launch director Mike Leinbach radioed the crew after they had been strapped in, the same as before, all over again. "Have a great flight and I hope you have a good turkey dinner packed for Thanksgiving."

"Thank you very much," Commander Wetherbee replied. "From the bridge of *Endeavour*, we're ready to set thundering sail."

At least they were going to get to try.

And in that moment something miraculous happened, the same miracle that always happens to the insides of astronauts. In that moment they go from being the only construction on earth more complex than their vehicle to the most simple. They become stones. There is no more thinking, no more emotion, no more remembering *Challenger*, no more wonder or dread. Instead, they slip into a kind of trance, quiet, serene, their minds wiped as clean as those of the last, brave members of a cult. The calm is a by-product of their years of training. It also springs from some small, remarkable part of them that they were born with. Mailer called it iron; Tom Wolfe called it the right stuff. But there is a lie in that poetry, because it makes that special something sound more exclusive than it is. The truth is, it's not just the dominion of astronauts. It's in all of us. There are millions of stories of ordinary people tapping it whenever they are trapped in extraordinary situations, whenever they might have otherwise seemed done for: when an engine on their plane's wing starts belching smoke, or when they're standing on a beach watching a hurricane blow in. Suddenly all that's left is their faith. Their bodies give them no other choice but to believe that everything might still work out, and, should it look like it will not— should things take a turn—next they find a way to believe that it was never meant to be. They think of everything that brought them to this moment, every step and side road in the history of their lives, and they see reason; they see, looking back at that long, crooked course, an artful conspiracy. Each of them comes to accept that in some profound way, we're all just passengers, and in the end, it's the universe that lives in us, not the other way around.

Like those millions of ordinary people, seven astronauts switched over to their own automatic pilots, leaving the worry for someone or something else to shoulder. All of the things that might have gone wrong or been mistaken—all of those parts, all of those hands—became remote, abstract, almost hypertheoretical. There were too many layers to sift through. There was too much for them to take in. And so they took in none of it. They settled back in their chairs, and they looked at their checklists, and they smiled to themselves. In the way that all of us will come to understand the facts of it, each of them already had: sometimes, our fates are no longer ours to decide, and we can only grip our fists until our knuckles turn white and hang on for the rest of the ride.

. . .

There was not a lot of conversation. The laughter and joking had stopped. The men of Expedition Six could follow along with their scripts, but they weren't to interrupt the rigid, technical dialogue flowing between upstairs and control. The only one of them with any sort of role was Bowersox, who, in the leftmost mid-deck seat, could reach the buttons that would allow him, in case of dire emergency, to open the hatch and deploy the wire-basket slide that would carry the crew, two at a time, down to the armored tank. If his services were required, more than likely they were seconds away from becoming corpses. Pettit, meditating beside him, and Budarin, at the far right, staring at his bee, tried not to lose their cool by thinking about that. But if only for an instant, all of them had needed to swallow their doubts, the bilious flutter of overwhelmed senses. How could they not?

On the one occasion when a shuttle crew had tried to subsume their nerves with idle chatter—the launch of STS-44—only one man had refused to take part in the charade. Story Musgrave, a four-time veteran of liftoff, had been tight-lipped in the middle of the dull roar. "Story, how come you're so quiet over there?" Tom Henricks, the pilot, had asked.

"Because I'm scared to death," Musgrave had replied.

The cabin had been nearly silent after that, like *Endeavour*'s was now, come the start of the aptly named terminal count, nine minutes from ignition. The main countdown had stopped cold there, the way it always does. The pause gives the test director time to call out a long list of acronyms, each representing one of the technicians sitting at a console dedicated to some small component of the shuttle and its launch. Each of them must respond to the roll call, but they have a limited selection of answers: one word (GO!) is good; two words (NO GO!) is bad. This time around, every one of them said a single word. Now only two of them could change their minds and stop the count. The supervisor of range operations continued monitoring whether any planes or ships had strayed into range of the splashdown. And the weatherman, fed data from around the world—the weatherman who had relayed the gloomy reports from Spain the night before—could still put the quit on things. But tonight he was as quiet as the crew, and the countdown was begun again, left to continue apace.

Seven minutes distant, hydraulic systems activated, and the White Room began to swing away from the shuttle, pulling back like a bomber creeping away from the charge he's just set.

Two minutes later, Wetherbee was given the order: "Go to start the APUs." The auxiliary power units provide the shuttle's hydraulic juice, and when they fired up, the crew knew that they were likely leaving. They were finally burning fuel, and burning fuel meant that only some very bad luck could stop them now.

After what felt like forever, counted down second by second—until they were just three minutes from launch—the shuttle's three main engines began gimbeling, testing their directional thrust. As they shifted up and down, left and right, throwing off a little push with each pull of the trigger, the astronauts could feel the shuttle swinging, like a skyscraper in a strong wind. It might have been unnerving were it not expected. They had waited so long for this moment, had imagined it so many times, now there were no surprises.

Thirty seconds later, they shut their helmet visors, and their oxygen came on. Each of them was now in his own universe. Each of them was under glass.

The flight data recorders switched on.

A little more than thirty seconds from liftoff, the shuttle's computers took over from the ground's. With each passing moment, another knot was untied, another set of handcuffs slipped. Bit by bit, they were being cut loose. Countdown, they had come to understand, is one long letting go. It's a goodbye filled with lingering until finally it's too late to turn back.

Ten seconds before launch, they heard the rumble of the water deluge system pouring out below them. The wall of water splashed into trenches carved out of the swamps, ready to catch their fire and dampen their acoustic shock. They blamed whatever trembling they felt in their hearts on the sudden burst. Everything around them had caught a bad case of the shakes.

Nine.

Eight.

Seven.

Six seconds from launch, the three main engines ignited. The cabin really began vibrating, the straps on the storage lockers swinging wildly. The shuttle's computers ran through a series of final checks, and every last one of them came back okay. Inside the crew cabin, Pettit strained to look out of the mid-deck's single porthole, a solitary five-inch-wide window back and to his left. All he could see was night, lit up with a glow.

Their headsets filled with noise, like static.

They could sense the shuttle pulling the stack toward its belly. It felt as though the beast was being held back, which it was, by those eight giant bolts, until it had built up the necessary thrust. It seemed like a long time for the crew to have to grit their teeth.

Five.

Four.

Three.

Two.

The solid rocket boosters kicked in. The bolts exploded.

One.

Now there was no going back.

Liftoff.

Almost instantaneously, they were pushed back into their seats by the force. Now they could decide how well the simulators had prepared them for the feeling. It has been likened to being strapped to the front of a freight train, or surfing the Loma Prieta earthquake and its aftershocks. Bowersox remembered it well. Budarin, accustomed to the bucking of the Russian *Soyuz* rocket, thought it was a relatively gentle shove. Pettit decided there wasn't any call for metaphors. To him, it felt exactly how he imagined it might. It felt like he was riding a rocket.

"GO WITH THROTTLE!"

And up they went.

Seven seconds into their flight, they cleared the tower, and the technicians in Texas took over from those in Florida.

More than thirty seconds later, the sound of their engines finally washed over the crowds gathered ten miles away in Titusville. Until then, hushed spectators had followed a silent light.

But inside *Endeavour*, it was loud. Engines roared. Equipment rattled. Everything shook. In the middle of chaos, there was nothing to do but wait, the idle members of the crew reminding themselves that every second that passed was one second less for something to go wrong.

After a little more than two minutes, they had reached an altitude of twenty-seven miles, and *Endeavour*'s solid rocket boosters were jettisoned, blown clear by explosives and eight small rocket motors. The crew let loose their first sigh of relief. They had outraced the ghosts of *Challenger*. Now if something went wrong, there were options: first Florida, then Spain, then California. Until then, there had been only go or no go, life or death.

It helped that their ride began to smooth out, even as their altitude and speed steadily increased. It felt less like they were in a rocket, and more like they were in the lead car of a very fast train.

Everything was normal. Everything was good.

Outside of their porthole was just more black.

Six minutes later, just eight and a half minutes into the flight, the main engines shut down, making the ride quieter still—until a loud clang signaled that the external fuel tank had been blown loose and begun to come apart, left to splash down, in pieces, in the Indian Ocean. The lost tank was like a penny thrown into a gorge from a bridge. It gave some gauge of the distance that the crew had covered, reminding them that they were a long, long fall from earth.

The reaction control system took over, tiny rockets that fired for short bursts, pushing the shuttle away from its fuel tank and making sure that it would eventually find its proper attitude, top down, belly up. Finally the shuttle slipped into orbit, at a speed of 17,489 miles per hour. Then, it was quiet. Then, the crew could breathe. They listened to the fans venting around them and the chatter of instruments and the best wishes and congratulations crackling up from the ground.

They felt lighter suddenly, as though they were lifting against their straps.

They could feel their shoulders relaxing.

Their jaws loosened.

And then, blinking away the sweat, Don Pettit caught something out of the corner of his eye. There, hovering in front of him, was Nikolai Budarin's bee. It was floating, weightless, tied down only by its string and looking as massive as a billboard: YOU ARE IN SPACE, it announced. That's when Pettit knew, despite everything that had happened, despite all that he had gone through—including his three heartbreaking rejections—that along with the others, he had finally made it. In eight spine-jarring minutes, he had become a real live astronaut. Along with six men, some New Mexican green chiles, and a toy bee, he had arrived.

2 THE OTHER SIDE OF THE ENVELOPE

From birth we're conditioned to accept the limitations of gravity. Only a lucky few of us can dunk a basketball; if we slip on a patch of ice, we'll land flat on our backs; whenever we drop a glass, it's doomed to smash into pieces. Suddenly, for the crew of *Endeavour*, none of these hard truths remained. Everything they had learned, everything they had come to expect of themselves and their environment in their time on earth, was no longer there for them to hold on to—nor did it hold on to them. One by one, they undid the straps that lassoed them to their chairs and began floating around the shuttle's cabin. For the rookies among them, Don Pettit included, the first few moments were a little clumsy; they looked like kids who had been thrown into the deep end of a pool for the first time. It would take them days, even weeks, to relearn how to do the things that on earth had been automatic, something as simple as moving from a seat to a storage locker, without looking as if they were flailing. Their lives had been stripped of their usual anchors, their leverage, their footholds and resistance. All that was left was a kind of lightness. They could feel everything, even their insides, trying to lift.

As is the case for many astronauts, the feeling made Pettit go green. On earth, everything in his body—half-digested food, mucus, stomach acid, waste—had been perpetually drawn down and out. In space, everything seemed determined to head up and away. It was as though someone had shouted "Fire!" in the crowded theater that was his guts, and now a packed house was scrambling for the exits.

But for the more experienced crew members, the feeling bordered on jubilation, as though they'd been cut loose somehow, un-shackled and relieved of their earthly burdens. For them, weightless-ness was freedom. Nikolai Budarin, who could seem hard-hearted and gruff on the ground, almost stereotypically Russian, smiled brightly and giggled, batting around his toy bee. He looked in for a nice, long high. Ken Bowersox, too, felt as though he was back where he most belonged, in a perfect, permanent state of flight. But for Pettit, the feeling was still foreign, even alien. He felt a little lost.

In a way, they all were. Buzz Aldrin, after he had bounced across the surface of the moon, talked about how much he had liked having something solid under his feet once again, even if it was only thick dust—how with the planting of that famous American flag, he had felt that he was *somewhere*. In deep space, he had felt rudder-less, just another nowhere man, and he had realized that it was gravity that he had been missing. Gravity, even just a little bit of it, had given him the feeling of being home.

Bowersox, Budarin, and Pettit were scheduled not to know that feeling again for a little less than four months—a semester in col-lege, football's regular season. Only minutes into their flight, they understood that from here on in, and in everything they did—eat-ing, working, sleeping, playing—there would be reminders that they were long gone. Now doing somersaults in the air, the first two looked ecstatic for it; the last hoped he would find his buoyancy soon enough.

Some astronauts took days to stop barfing, and did so only af-ter injecting themselves with buckets of Phenergan; others were left chewing handfuls of aspirin, combating the aches in their legs and spines, which can begin to stretch out and lengthen as much as an inch without gravity closing the gaps. Already, Pettit felt as though he were on a rack.

Before launch, he had received counsel from a classmate who had beaten him into space. As well as offering a few salient tips on how to use the toilet—something along the lines of the best defense is a good offense—he had told Pettit that he might want to have his first ten minutes in orbit rehearsed, blocked out the way stage actors waltz through a play. Otherwise the experience and discomfort would overwhelm the necessities of it. No time was allotted for breath-catching. There was so much work to be done.

Pettit had taken in the advice. He had memorized a short, opening to-do list down to its tiniest detail, and now he began working through it, and his upset stomach, step by step. First, he raised his visor. Next, he took off his gloves and, after losing hold of them for a split second, he pushed them under a strap that was wrapped around his knee, making sure that they couldn't float away. Then he popped off his helmet and his headset, tucking the second inside the first. He churned his way toward the rows of storage lockers and found the bag that had been earmarked for his gear, labeled not with his name but with his mission designation: MS5. He slipped the helmet, the headset, and his gloves inside the bag, sealed it up, and put it away. As though he was following a recipe, he continued the process until he was down to a comfortable set of clothes that he might have worn to the gym. Come the end of the personal dismantling, he still felt sick, but at least he had made it through the opening act without a stumble.

The rest of the crew had also freed themselves from their bulk, and now they began working through checklists and itineraries dictated by the ground, timed down to the minute. For the first few hours in orbit—"post-insertion," it's called—the crew's cabin is abuzz. For Expedition Six, still isolated down below on mid-deck, their principal job was to turn their cold, sterile surroundings into something that looked more like a train's sleeping car. They folded away the seats, activated the cooling system, set up the exercise machines, fired up the galley, and rolled out their sleeping bags. It was like those frantic few hours after you pull into a campsite, and the tent needs to be pitched and wood needs to be gathered before dark. Unrolling the sleeping bags gave each of them that feeling especially.

But this was no ordinary wilderness they were in. After about ninety minutes of yeoman effort, Pettit had lost himself and his sickness in the hurry. He had maybe even forgotten where he was, exactly, at least until he visited the cockpit for the first time. He laughed when instinct still made him grab hold of the ladder that pointed the way, even though it was now about as useful as a paperweight. And then, with the space shuttle turned upside down, and with two large windows making up most of the cockpit's ceiling, he floated into what felt like a bubble of Heaven. There, filling his eyes with a warm, soft light, was a panoramic view of earth, of white clouds and blue ocean, half of it bright with sun and half of it dark with night.

He blinked once, twice, three times. For maybe ten seconds, he stopped, losing hold of everything he had left to do, instead caught stealing a look at everything else that had opened up for him. A big part of him wanted so badly to press his nose against the glass, but the rational rest of him—not to mention Bowersox and Budarin, still plugging away—called him back to work. He pulled himself from the window, telling himself that he would have all of the time in the world for staring.

The rest of the trip passed as though in a movie, projected on the giant blue screen stretched out underneath them. Everybody took his turn in a front-row seat. They could see the Blue Nile meeting the White Nile at Khartoum, boat wakes and vapor trails, the northern and southern lights in the same shift, cloud shadows that stretched for hundreds of miles.

The rapture was tempered only by the still-haunting specter of absent friends. During *Endeavour*'s long chase of the International Space Station—it would take nearly forty-eight hours to track it down—Bowersox gave a surprisingly revealing interview about Don Thomas to the Associated Press. Maybe, as with the bee, there were still strings attached after all.

"This emblem that's on our shirt was designed by Don," Bowersox began, talking about the triangular Expedition Six patch that had been stitched to their flight suits, depicting a stylized station eclipsing the sun during yet another orbit around the earth. "So he's with us every minute in spirit and we think about him a lot and we can only wish him the best. We know this has been very, very hard for him, so that's been the toughest thing for us, too. But he's a big part of this mission. Everywhere we go we see reminders of him, and there's no way we could not think about him."

At the same time, Bowersox admitted that he hadn't spoken to

Thomas since that awful night when he told his friend that he had been bumped from the mission: "It's still kind of painful and sore for Don. When he talks with us, it becomes even more painful. We're going to try to connect with him when we get on orbit or after we get home, after he's had a little bit of distance. But this is a hard thing, to be so close to accomplishing a dream—going for a long-duration flight was Don Thomas's dream—and when he wasn't able to do it, it hurt him pretty bad. So as the distance and the time heals that wound, then I think it'll be a little bit easier for him to discuss how much fun we're having in orbit. I'm happy to get to fly with Don Pettit, but I was really looking forward to flying with Don Thomas, too, because he's such a great guy."

Like athletes, however, the men of Expedition Six needed to start thinking only of tomorrow. Along with the rest of *Endeavour*'s crew, they helped examine the Canadarm to make sure its earlier brush with disaster really hadn't limited its function. Happily, it appeared to work just fine. Testing it further, the crew turned on the cameras attached to the end of the arm and began scanning the luggage stowed in the shuttle's massive payload bay. The biggest, most important piece of cargo would soon be locked onto the International Space Station: the P1 truss, a $390 million chunk of station's ever-expanding backbone, a girder assembly that all on its own measured forty-five feet long. In addition to supplying structural support for the modules that housed the astronauts—acting like the sticks in a giant kite or the flying buttresses in a Gothic cathedral, depending on your feelings toward station—the truss contained three ammonia-filled radiators, folded up like accordion panels, that would help vent the excess heat generated along with much-needed electricity by the station's solar panels.

Looking at the almost ghostly images of the truss from the safety of the crew cabin, Expedition Six reveled in their special delivery. They could see something beautiful through its utilitarian mass—not so much in what it was but in what it represented. It was a part of what has been called the most ambitious single construction project since the Great Pyramids, and now it had made the giant leaps from drawing board to space to impending installation.

With it, Expedition Six was about to become a part of history, too.

. . .

After two nights in space—sleeping with their heads strapped to their pillows, so they could feel the softness of them, and with their arms sometimes floating freely, leaving them looking like drowned sailors—*Endeavour's* crew woke up excited to catch their first glimpse of their final destination. With each successive rocket burn, still being coordinated and fired from the ground, the shuttle was drawn closer to what first appeared as a small, white light, like a star that was brighter and closer than the rest. Inside that light, Expedition Five—two cosmonauts, Valery Korzun and Sergei Treschev, and the American Peggy Whitson—waited eagerly to greet their visitors. Because of *Endeavour's* earlier troubles, they had spent nearly 170 days in space, and they were ready for new faces and hugs.

"You guys look pretty good out there," Whitson radioed.

"We were just saying the same about you," Mike Lopez-Alegria said in return.

Although both the shuttle and the station each rocketed across space at five miles a second, their relative distance closed slowly, and for good reason. There was disaster lurking behind their dance. A collision would almost certainly see these ten astronauts made to look more permanently like drowned sailors, with their newfangled ships done in by old-fashioned holes.

Nine miles out from station, at 2:15 on a Monday afternoon, a short burn of the shuttle's left-hand orbital maneuvering system put it on a nearly perfect course. At that point, the ground began ceding some of its control, giving permission for Commander Jim Wetherbee to look through his window and fire as many as four short correction burns over the next couple of hours. Once *Endeavour* was within six hundred feet of station, the previously shared load was placed entirely on him. With a gentle touch of the stick, he could nudge the shuttle up or down, left or right, as though lining up the biggest pool shot of his life, until he was satisfied that he had found his dead aim at the docking port in Destiny, the last in the station's chain of modules. When it came right down to it—having come so far, so fast—Wetherbee had a margin of error of just three inches to work with. Not surprisingly, he took his sweet time in closing the last of the gap.

Docking was scheduled for 4:26 p.m., but it wasn't until just before five o'clock when the shuttle and the station finally connected, 250 miles above the great blue expanse of the South Pacific, making contact just behind the top of the shuttle's cockpit. For the last several feet, Wetherbee had slowed his approach down to a little more than one inch per second, a veritable crawl considering the urgency with which the shuttle had left earth. The union was as soft as a first kiss.

The astronauts waited for the gentle waves of their impact to subside before they triggered the hooks and latches that would keep the shuttle in place. It took another hour for the small tunnel that had formed between the shuttle and station to pressurize. It was checked for leaks, and after they had been given the all-clear by the ground and each other, Expedition Five and *Endeavour*'s crew opened their respective hatches and finally, literally, flew into one another's arms.

With Whitson snapping pictures, Wetherbee was the first to cross the threshold into station. Bowersox followed closely behind, already in the blue shorts and sock feet that he would sport for most of his mission, and next came Budarin, still giddy and smiling. Hanging like subway riders on the restraint bars bolted to the station's ceiling, everybody hugged and clasped hands with everybody else and started shouting greetings in Russian and English at once.

"Nice to see you!"

"Nice to see you, too!"

But with only Wetherbee, Bowersox, and Budarin having made it on board, the flow of traffic stopped. "Where is everybody?" Bowersox asked after planting Whitson with a peck on the cheek. At last, Pettit burst through the hatch, bearing two warm silver bags with straws jammed into their sides.

"Coffee!" Pettit shouted, handing the bags to a laughing Korzun and Treschev. The Russian crewmates had each run out of caffeine during their extended mission and made a desperate pitch for some more. Pettit had obliged by sharing two servings out of his private, under-the-seat supply. "Coffee!" he shouted again, nearly doing a backflip on his way through Destiny.

It would be a while yet before he could explore the rest of his new surroundings, although he, Bowersox, and Budarin were immediately impressed by the size of their new home, how roomy it seemed after they had spent nearly two full days packed into the shuttle's claustrophobic quarters. It felt as though they had stepped into the foyer of a mansion, or at least the makings of one. Like Pettit and the happy rush he had experienced when he first saw Budarin's toy bee take flight, the three men were flush with feelings of arrival, sharing the relief that comes at the end of such a long journey.

Relief, but not rest. Aside from unloading tons of supplies from *Endeavour* and trying to make themselves feel a little more at home—Expedition Six took over the station's three sleeping compartments, about the size of phone booths, from Expedition Five on their first night on board—they also had to pitch in during the installation of the P1 truss, three days of work scheduled to begin the next morning, Tuesday.

Wetherbee, at the controls of the Canadarm, continued his busy assignment by lifting the truss out of the shuttle's payload bay. Only his grasp kept it from floating into space, and there it dangled, against the black, looking for all its size like a good sneeze might blow it to the edge of the universe. But by using every inch of the Canadarm's fifty-foot length, Wetherbee was able to bring the truss safely within reach of the station's own Canadarm, operated by Whitson from inside Destiny. Each kept hold for nearly nine minutes, just to make sure that the handover was true. When Wetherbee let go, Whitson was able to swing the truss into its final position, aligning it end to end with the already installed S0 truss, the station's primary vertebrae. It fit together seamlessly. Using the computers inside station, Pettit commanded a claw on one truss to grab a bar on the other, tacking it in place. Now the connection had to be made complete.

Lopez-Alegria and John Herrington had spent most of their day getting ready to head outside. They had already pulled on their white spacesuits, climbed into one of the station's airlocks, let the oxygen out, and opened the hatch. Now they took a deep breath and leaped out into the darkness. With the earth spinning fast beneath their feet (a sight that had the unnerving effect of making them feel every so often as though they were falling), they first worked at connecting the P1 truss's power, data, and fluid lines, critical if its girders were to keep from freezing.

Also attached to the truss like a parasite was something called the Crew and Equipment Translation Aid (CETA), a kind of flatbed mining cart that, in the future, would run along rails stretched the length of the completed truss. (It was designed to help spacewalkers lug heavy equipment from one side of station to the other.) The cart had been locked into place to keep it from moving during the shuttle's flight and the truss's installation, and now those locks needed to be released. While they were at it, Herrington and Lopez-Alegria removed some of the pinlike hardware that had kept the truss itself in place during launch.

Lastly, they attached an antenna that would pass along signals from the cameras that were sometimes strapped to the side of an astronaut's helmet, especially when he had been ordered outside. Already those cameras had provided some of the most terrific images of earth: when the view cut from an astronaut's gloved hand, tightening a bolt with an ordinary wrench, to the whole of Australia, say, baked golden brown under a hot sun, it was almost hard to make sense of the scene, the mundane and the spectacular suddenly side by side. It was like watching an electrician connecting a wire and flicking a breaker that somehow turned on the moon.

But up there, for them, it was starting to feel like just one more job that needed doing—not routine, but on glamour's wide spectrum, more routine than red carpet. Astronauts have a higher threshold for drama than most of the rest of us: after you've survived the thrill of a shuttle launch and snored in space for a couple of nights, your sense of perspective takes a hit, the way a near-death experience might make you care less about who's going to win the

American League. And so, after taking a day to recharge their spacesuits as well as their own batteries—and to celebrate Wetherbee's fiftieth birthday, which earned him some ribbing from the ground—Herrington and Lopez-Alegria headed outside once again. Like shift workers punching the clock, they picked up where they had left off, releasing the launch locks on the three radiators and pulling the CETA cart toward its home position, wedged onto the S0 truss.

After another daylong break (although packing and unpacking continued apace), Saturday called for more of the same, finishing up the truss's installation with six hours of plumbing. Exhausted and fresh out of adrenaline, Herrington and Lopez-Alegria clambered back inside and slept as soundly as men who had seen everything in their lives go exactly as they had hoped it might. Making full use of their eye masks and earplugs, they slept a deep, dreamless sleep.

When they woke up, the last morning's worth of work was well on its way to being finished. A computer printer, sent up to replace a balky one that had already been punted into the station's junk drawer, was the last bit of cargo brought on board. Expedition Five had nearly finished their packing up, too, stowing the last of their gear in *Endeavour* for the flight home. As a goodbye gift, the ground had given them and their colleagues the rest of Sunday afternoon off, mostly to prepare themselves for Monday's departure. After the frenzy of the previous few days, it was a much-needed chance for everybody to say goodbye—three to the station, and three to the earth.

. . .

Part of the farewell process included formal exercises, usually orchestrated by the ground and held in front of cameras. Among these was an elaborate change of command ceremony, which concluded with the following exchange between station commanders old and new:

"Ken, I'm ready to be relieved," Valery Korzun said before passing the radio to Ken Bowersox.

"I relieve you of your command," he said.

"I stand relieved," Korzun replied.

As terse—and rehearsed—as the conclusion might have appeared from the vantage point of living-room couches, for the people involved (especially for Expedition Five, about to abandon the only home they had known for nearly six months), it was an important part of the leaving ritual, as symbolic and poignant as a torch going out.

But more important, perhaps, were the quieter, more private moments before departure. There were meals enjoyed together, and last, longing looks taken through the windows, the bundling up of photographs and stowaway keepsakes. That was the hardest part, because what once were treasures suddenly seemed disposable, like the astronauts themselves, having served their purpose and been told that it was time for them to go.

By Monday morning, the feeling had forced Korzun, Treschev, and Whitson into making a subtle mental switch, the building of a distance between themselves and station. Each of them would have happily stayed longer if needed; a small part of each of them even wanted to—the same secret part of them that had felt stung when they were first overrun by the new arrivals. But now they turned their minds from their leaving one home to their imminent arrival at another. (For them, for all of us, it has always been easier to swallow the idea of saying hello rather than goodbye.)

Korzun, a veteran of a long-duration mission on Russia's Mir, braced himself for the assault of gravity, imagining himself stepping down from the shuttle's cockpit onto solid ground. Whitson gave in to lighter-hearted fantasy: she began telling everyone how much she was looking forward to a thick juicy steak, a Caesar salad (with lots of garlic), and a Coke drunk from a glass filled with ice.

Korzun had turned his return into a test that he looked forward to passing. Whitson had turned earth into a resort. In their own ways, both had found something to look forward to when they finally shared a last round of hugs, floated into the shuttle, and closed the hatch behind them. They had found something to soften the blow.

. . .

Inside station, suddenly it felt as though a big party had finished and the revelers had emptied out, and all that was left in their wake was quiet and mess. Expedition Six had looked forward to this moment from their first seconds on board. Until that hatch had shut, each of them had worried that there might be a malfunction or a change of plans or some internal mutiny that would have seen their time on station end before it really began. Until the shuttle had undocked and disappeared from view—until Whitson was left looking back tearfully at her old home, having forgotten how beautiful it looked when the sun's rays struck it, flashing like lightning—there was always that chance, however remote, that Bowersox, Budarin, and Pettit would be remembered as visitors instead of residents. Now, station was theirs, and they were station's.

Even as tightly packed as they were, together the men of Expedition Six were more alone than they had ever been in their lives, more alone than most of us could ever even imagine being, adrift in the middle of a vacuum, almost 250 miles above the surface of the earth. For the next fourteen weeks, they could drop in on their families and the rest of the world by Internet phone, radio, and e-mail (they could even order flowers for their wives if they woke up feeling frisky), but if they wanted for genuine company, they would have only one another and the insides of this giant machine to turn to.

In time, they would come to know the International Space Station as well as anywhere (or anyone) waiting for them back in Houston. In time, they would grow to think of it as a living, breathing thing. They would learn its secrets, and they would be able to close their eyes and trace every detail of it with their fingers in the air in front of them. But in those first few moments after their friends had left them behind, they were strangers to the only place in the universe they had to call home.

At the Johnson Space Center in Houston, there's a full-size mock-up of the station's interior, and Bowersox, Budarin, and Pettit had spent some time in it, feeling it out, but really, it was the least effective of their simulations. Those hours had given them a sense of the scale of the place—inside, it is about 150 feet long and eight feet wide—but little else. Apart from a single, empty sleeping compartment, the mock-up had been undressed. The walls were lined with mural-sized photographs of the insides of the real thing, but there were no knobs or switches, no tubes or dials. There was none of the persistent background noise, no whirring fans or computers, no clicks or groans. There were no cables to get tangled up in, no bulkheads to bump into, none of life's debris. Everything was smooth and clean, all perfect right angles; pacing off its length felt a lot like strolling down a brightly lit hallway with no doors.

Now there were still no doors. But neither were there floors or ceilings, nor was there any up or down. There was only the reasonable facsimile of walls, and just about every square inch of them was occupied with tools, spare parts, boxes of food, control panels, cameras, laptops, racks of scientific experiments, and personal effects—Nike sneakers bound behind elastic straps and photographs taped up, as well as sleeping bags, toothbrushes, shaving mirrors, utensils, portable compact disc players, an Australian didgeridoo (owned by Pettit), and one really ugly necktie (brought up by Bowersox).

Floating their way through this cluttered, closed-in tunnel, they now felt like actors on the most elaborate set ever built, put together by hundreds, even thousands, of designers, builders, and prop masters. Station had been in orbit just long enough for it to feel experienced; its sharpest edges had been worn down and its fresh-from-the-factory luster had been scratched and tarnished. But it was also new enough for it to feel as if it was still being built, even invented—which, of course, it was. As with a house stuck in the middle of a perpetual renovation, there was a kind of sketched-out order to things, but there was also plenty of chaos and dust. Until the last module had been dreamed up and launched, until the final piece of the puzzle had been put into place, it would always feel as though the workers had gone home for the night but were ready to come back and pick things up again tomorrow, having left their chop saw on the kitchen table in the meantime.

When the International Space Station is finally finished, it will look like an enormous mobile, revolving around an axis of modules—silver, vaguely cylindrical rooms bolted to each other mostly end to end—built and equipped by one or more of sixteen countries: principally the United States and Russia, but also Canada, Japan, the eleven nations of the European Space Agency, and Brazil. Although the station's blueprints have been continually scribbled over and updated since its inception, the long list of planned modules includes at least six laboratories; a couple of dedicated living spaces with room for as many as seven astronauts; a checklist of nodes, adapters, and docking ports; and an ever-ready *Soyuz* capsule, Russia's age-old means of space exploration, now fulfilling the role of lifeboat. The entire structure, running perpendicular to the backbone of trusses that *Endeavour*'s crew had just added on to, will be powered by almost an acre's worth of solar panels. And at an estimated final dimension of 356 feet across and 290 feet long (weighing in at more than one million pounds), the completed station will easily eclipse the next largest manmade object ever in space. It will be more than four times as large as Mir, the world's first multimodule space station. It will look as though we've slipped another star into the sky.

But in November 2002, the International Space Station was still a long way from throwing off its own light. Only three principal modules had been strung together, making it look more like a tincan telephone than one of the wonders of the modern world, the whole of it electrified by just eight slim solar panels. The truth of it was, Expedition Six had moved into a home that was more dream than reality, and Bowersox, Budarin, and Pettit were left trying to make themselves comfortable in a basement that had just been poured.

· · ·

At one end was Destiny, the scientific laboratory built and financed by the Americans. It was also one of the newest modules, launched less than two years earlier, in February 2001. Where the shuttle had been docked, there was now a closed hatch. Sometime in the future, it might be propped open, leading the way to another, as yet absent module. But for Expedition Six, it was a portal only to nothingness.

Turning their backs to it, they saw a long, narrow space about the size of a school bus, packed with a tangle of power cables, computers, panels, and scientific gear. For most of us, it would seem cramped to the brink of collapse. For someone like Don Pettit, it was the only playground in the universe that stood a chance at satisfying his insatiable curiosity.

He liked it so much that he slept in it, in the most isolated of the station's three sleeping compartments. Inside, a sleeping bag had been strapped to one of the walls. (When the station is complete, this particular sleeping compartment will be moved to one of the proposed habitation modules; for now, it was just the most convenient place to stow another body, among the racks of plants, fluids, combustibles, and medical paraphernalia.)

Across from and next to Pettit's bedroom, there was space for as many as thirteen racks of experiments. Running along the module's two "sides," they helped manufacture an artificial sense of ceiling and floor—with top and bottom further defined by Destiny's few patches of white space and hand- and footholds painted electric blue. They helped the astronauts orient themselves in a world otherwise without landmarks. The racks had the added advantage of being interchangeable, their function dependent on the interests of the current residents and the wishes of government and civilian scientists on the ground. Each of the experiments had inaccessible names like the Fundamental Biology Habitat Holding Rack and the Materials Science Research Rack, but really they were just the boxes in which the rats were kept, or where dishes of proteins and crystals were seeded.

Jumping out from the clutter was also something called the Microgravity Glovebox, which Expedition Five had installed and quickly broken. Like an incubator for a premature baby, it was a sealed container with built-in gloves, allowing the crew to handle more exotic experiments—those involving fire, say, or gases—without having to worry about poisoning themselves or turning their home into a nebula. Whenever Pettit floated past this dormant, precious thing, he let out a little sigh, hearing the echoes of the stern orders from the ground: DO NOT TOUCH. He wondered how they

could ever find out if he ignored their demands—and, more important, how they might punish him if he did. That broken glove box
gave him the resolve to earn his way into a longer leash.

In the meantime, Destiny offered plenty more to occupy his
time. In addition to the bundles of experiments, the module housed
a series of IBM ThinkPads that monitored and controlled the health
of station. (Rather than running things with old-timey switches and
buttons, Bowersox, Budarin, and Pettit could issue commands
through on-screen tabs.) There was also the joystick that was wired
into the robotic arm, as well as the means to fire up cameras that
could scan the station's exterior and project the images on a collection of monitors, making it look like the security guard's desk in the
lobby of a high-rise.

But Destiny's architectural highlight was a huge, beautiful, circular window near the center of the module, surrounded by rows
and rows of silver bolts. (A break in that seal would release a high-
pitched, ultimately heart-stopping whistle.) Almost two feet wide
and an optical gem, that window offered the clearest views of earth
from station. More often than not, someone was next to it, unblink-
ing, open-mouthed. Through its perfect glass, astronauts had taken
pictures of avalanches and plankton blooms that looked like
high art.

. . .

At the end opposite to the hatch, Destiny tapered into an open-
ended cone, through which Expedition Six kicked into a small space
called the Node, or Unity, because it helped bridge the American
segment and whatever modules Russia chose to send up—Zarya
and Zvezda, thus far. In some ways, it was a neglected space, more
of a way to get somewhere else than a place of its own. But for some
of the station's past crews, and soon for Bowersox, it became a favorite sanctuary. There wasn't much in it: a few large white bags
filled with laundry and supplies, a wall of water-filled containers,
and on the "ceiling" there was a resistive exercise device, on which
the crew could do upside-down squats using giant rubber bands.
But there was something more to it. Whether by accident or by de-

sign, it was somehow more welcoming than the rest of station, warmer, more homelike. It was a kind of cocoon. The lighting was soft, and it was usually quiet, and its interior had been painted a different color from the rest of station's stark white—almost a shade of pink, closer to coral. Astronauts who hadn't visited the International Space Station sometimes poked fun at the Node's anonymous interior decorator and his burst of wacky flair, but those who had been there expressed an appreciation for it that bordered on love.

Jim Voss, who had spent 163 days in station with Expedition Two (and who had spent parts of many of those days tucked away inside the Node), summed up those good feelings best: "I had always laughed when they talked about using that color for the Node interior," he had said after his return. "I thought, Who came up with this? But whoever it was did a wonderful job, because it made it the most pleasant place. That was where I went whenever I wasn't able to sleep, or when I wanted to be alone to listen to music."

It would become the same sort of retreat for Bowersox; he thought of it as a place in which he could unwind, dream, relax. It had become the station's Finnish sauna.

Off to one side of it was another small space, but one that wasn't the scene of much easy breathing. The Quest Airlock was the jumping off point for American space walks. In it, the torsos of two bulky white spacesuits were stored, lined up like powered-down robots. And beyond them, a large hatch waited for the next time someone would open it up and see nothing but black.

Back into the Node, Bowersox, Budarin, and Pettit passed through a narrow tunnel, two more open-ended cones bolted together. It was made to feel tighter for all of the supplies stored in it, as well as a couple of fat white ventilation tubes that ran through it. Down through its floor was another docking port, one that the Russian *Soyuz* capsule could latch on to. And just a little farther along, Zarya (the Russian word for *sunrise*) opened up.

Zarya would always hold a certain distinction for being the International Space Station's first segment, slung into a solitary orbit on an unmanned rocket in November 1998. But it never had much more going for it, including a technical name about as unromantic as it gets: Functional Cargo Block. Boiled down, the module was little more than a big closet, glossed up by two small solar panels that extended out into space from either side of its hull. Its four interior walls were filled with boxes of food strapped down by bungee cords, along with more containers of water and dozens of batteries. It also housed an amateur radio, marked by a set of headphones clipped to the wall. With them, Expedition Six could listen in to whatever eager ham radio operators they happened to pass over on their ever-changing orbits. But more often, they would use Zarya for more mundane purposes, like hanging out their laundry to dry. There were usually at least a few pairs of shorts and a couple of T-shirts pinned down by their waistbands or a single sleeve, the rest of them floating in the air, stiff-looking, as though they had been given a nearly lethal dose of starch.

What Zarya lacked in charisma, Zvezda (suitably, Russian for *star*) made up for. If Destiny was the station's brain and Zarya was its whiffy armpit, Zvezda was its heart, without a doubt.

For mid-deck-bound Expedition Six, entering this, the largest of the station's modules, was like opening the flaps of a circus tent. Also called the Russian Service Module, Zvezda was forty-eight feet long and airy, peppered with fourteen windows. Relative to Destiny, at least, it was also more shipshape, with most of its purely functional equipment having been locked away behind panels. The spareness hid Zvezda's true importance to station. With its flight control and propulsion systems, its occasional rocket firings helped keep Expedition Six in orbit. With its galley, toilet, exercise equipment, and sleeping compartments, it helped them meet their more human needs, too. Given the magic inherent in spaceflight, sometimes it's easy to forget that astronauts, as otherworldly as they might seem, still need to sleep, eat, and go potty. This was the place for it.

In the main, Zvezda was a near replica of Mir's principal chamber, now more than twenty years old. Along its walls were some of the expected pieces of computer hardware and a collection of restraint bars, but behind its many panels were machines of a more honed, necessary grace: a system that takes water out of the air and purifies it enough to drink; another that zaps that same water with charges of electricity, breaking it down into clean, fresh hydrogen and oxygen. (The water recycling was particularly important: without it, shipping up enough H_2O to sustain the station's crew would have cost the program $700 million each year.) But like the insides of a watch, most of its complex mechanics were hidden away behind a more attractive face, and for the men of Expedition Six, Zvezda was like a loft apartment without the exposed brick.

Of its luxuries, the most important was the galley, calling out to Bowersox, Budarin, and Pettit from the starboard side of station, toward the far end of Zvezda. Around and behind a red Formica table with plenty of foot restraints bolted underneath it—so that the crew could sit down and eat dinner like civilized people—there were hot and cold (actually lukewarm) water taps, a food warmer the size of a briefcase, and dozens of metal-faced drawers filled with meals, snacks, and drink bags. Like kitchens on earth, it was a busy, social place. It didn't hurt that a surprisingly good sound system had been hooked up, helping boost the astronauts through their daily routine. During mealtime, Expedition Six would crank one of the several albums that had been brought up to station and left over time. Santana proved a particular favorite. Weightlessness made for some terrific air guitar.

On the floor next to the table sat a treadmill, on which a strapped-down Bowersox, especially, and again accompanied by music, liked to run for daylight. On either side of the treadmill, two sleeping compartments faced each other, Budarin's starboard, the galley side, and Bowersox's port. Like Pettit's pad back in Destiny, each contained a sleeping bag unrolled against an otherwise empty wall, a laptop computer, and a small window. Come lights-out, it was a little like slipping into one of those Japanese hotels that rents out human-size mail slots rather than genuine rooms, but still, they were private and cozy, havens that provided just enough personal space to keep a man from cracking up.

For Bowersox, however, Zvezda was noisy, noisy enough to disturb his sleep. In the end, he would occupy his designated bunk for about a month, just long enough to make sure that the Russians weren't offended by his moving out. He would end up floating his sleeping bag into his beloved Node, where he would spend the rest of the mission snoring with his back against water bottles. He made that call not just because they felt a little like a bed, but because water is an excellent radiation shield, and try as he might to block it, Don Thomas's fate had never been far from his mind.

Bowersox hadn't liked sleeping next to the toilet, either, shining just beyond Zvezda's sleeping compartments. It was a tiny, square, metal contraption that, when its lid was closed, looked like it might make waffles.

Going to the bathroom has always been one of the great perils of space exploration. For men, anyway, urinating is usually easy enough, at least with the help of a hose and some suction. Taking a crap is the bigger pickle. The most horrific stories have been passed down among astronauts like the kinds of fables that make children scared to look under their beds. Through them, veterans bestow the lessons they've learned with gravitas; rookies, however, will still no doubt find themselves in pitched battle against a wayward shit. The goriest details are usually kept within the astronaut fold—deemed too graphic and uncomfortable to share in polite (read: civilian) company, like the persistent rumors of space sex. But the bottom line is that in space, whatever isn't pinned down takes flight, including poop. In other words, bowel evacuation in orbit is never a passive exercise—a certain velocity must be given to the offending projectile, enough to deliver it to the surface of the toilet's bowl and make it stick. It's just too bad that sometimes there isn't enough gunpowder in the cannon.

It was too bad, too, that the station was still without a shower. (It won't arrive until one of the planned habitation modules makes the trip up.) Because water remains a scarce resource, crews have had to make do with a quick splash and a scrub-down with moist towelettes. Hence, there is a particular urgency when it comes to making a clean getaway from the can. For some astronauts, it's enough to turn their guts into concrete.

What goes in must eventually come out, however, a harsh reality in realms beyond the toilet—more specifically, in the end cone that caps Zvezda. Occasionally it served as a docking port for *Progress*—an unmanned Russian cargo ship used to ferry loads of food, clothing, spare parts, letters, and small gifts from home into the crew's open arms—but most of the time, the hatch was filled with garbage, waiting to be taken out to the curb.

With Expedition Five having bundled up their crushed cans and empty foil wrappers and taken them back down to earth, the passageway was clear, but that tidiness would last for only a short while. It didn't take long for Expedition Six to begin filling up the station's empty spaces, and not just with their trash. The men and women who had gone up before them had already left their own legacies, each one of their contributions layered on top of the last, like so much graffiti on a once white train. There was a silver ship's bell mounted on a bracket in Destiny, across from Pettit's sleeping compartment; a photograph, stuck on a bulkhead above the galley, of a white-bearded Konstantin Tsiolkovsky, the father of modern space flight; and enough musical instruments to put on an impromptu concert, including an electronic keyboard and an acoustic guitar.

Now it was up to Bowersox, Budarin, and Pettit to make their own marks on station, to leave their own fingerprints, to find their own memories. It was up to them to furnish their new home, and to make it theirs alone.

. . .

It wasn't always a conscious acquisition. There was the usual course of arcane experiments to run (most of them involved measuring the effects of weightlessness on various materials), and there was a demanding maintenance program that had to be kept up (batteries needed charging, computers needed rebooting, filters needed unclogging), but Expedition Six's principal assignment was simply learning how to stay alive in space.

For its truest believers, the International Space Station has always been an outpost, a stepping-stone, a pit stop. It is the means to learn how to live longer and better in space, and, someday, much farther away. It will take two years to travel to Mars and back, and before men can make that kind of trip, they first have to master less ambitious journeys; before miracles, there are errands to be run. Astronauts have to spend one, two, three, four months and more on station, testing their physical and psychological limits, learning like free divers how to push deeper and deeper each time out. Ultimately, they will need to find a way to live in space so happily that they will start to forget that they have ever missed earth, their days filled with the fundamentals of a new, weightless life. Expedition Six was charged with learning again how to brush their teeth, cut their hair, wash their clothes, the boring things that help anchor those magical moments that make life really worth living. And if they managed to do each of those boring things well enough, simply enough, Bowersox, Budarin, and Pettit would have found an astronaut's nirvana. They would have learned how to push through to the other side of the envelope.

. . .

Their quest began first with breakfast, then lunch, then dinner. In their daily schedule, eating occupied much of their waking life. Each member of Expedition Six had a personal menu (although Pettit's remained tailored to Don Thomas's culinary wishes), built into that strict eight-day rotation. At the end of the cycle, they would go back to the beginning and start plowing through it again. It didn't help that, because of their general state of zero-gravity congestion—their faces swollen with fluid and sinuses blocked—food tasted blander than it might otherwise. There are stories of shuttle astronauts having filled their two-week-long menus with the same few meals, indulging in stacks of their favorite foods, only to crack open the first pouch and find something inside that tasted like puke. Variety, then, is important. So, too, is making a couple of zippy selections, such as shrimp cocktail (probably the most popular food in space because of the horseradish) and spaghetti with spicy meatballs. Still, Pettit remained grateful for his cans of green chiles, which he sprinkled on just about everything he ate. He had opened his first can with his crewmates, who wrinkled their noses at the acquired taste. Thus freed from the social obligation of sharing, Pettit guarded his little supply as if they were gold coins, recalling the food wars on STS-40, when taco sauce became currency, traded for favors and hidden away in secret stashes. The astronauts on board that flight even drizzled taco sauce on their Rice Krispies in the morning, just so they would have something to taste.

For astronauts tucked away inside station, food selection is even more critical—not just for flavor but to fight their body's inevitable decay. They can't go off their appetites; they can't get caught up in dreams about a steaming hot pizza fresh from the box or Peggy Whitson's steak and Caesar salad. Each day, they have to push through all that they have to choose from, three nutrient-packed meals with plenty of snacks in between.

In the case of Expedition Six, NASA's food laboratory had balanced their diets between six categories: Beverage (B), Rehydratable (R), Intermediate Moisture (IM), Thermostabilized (T), Irradiated (I), and the rare, blissful Natural Form (NF). (Once in space, the food was divided by a more mouthwatering nomenclature, including "Vegetables, Soups, and Sides" and "Snacks, Sweets, and Yogurts.") Because of space and weight limitations, not a lot was included in the way you might pull it off a grocery store shelf. In Nikolai Budarin's case, crackers, buns, cookies, nuts, cinnamon rolls, and hard chocolate were about the only things he could eat without some kind of preparation, whether it was adding water to Day 7's breakfast of Buckwheat Gruel (R), heating up Day 5's always mysterious "Appetizing Appetizer" (T), or hauling into Day 4's nuclear-fired BBQ Brisket (I). But it wasn't as bad as it sounds. Each day boasted a thoughtful mix of Russian and American munchies with the occasional dash of more exotic inspiration, including Kharcho Mutton Soup, Peach Ambrosia, Pork with Lecho Sauce, and Jellied Pike Perch. Budarin had also fortified his menu with staples that he couldn't do without and enjoyed almost daily: tea with lemon and sugar, gallons of apricot juice, and prunes stuffed with nuts. (They came in especially handy when Day 1's thermostabilized meat loaf got caught in his pipes.)

For all of them, but especially for Pettit, mealtime became a chance to experiment as well as to refuel. Eating became an elaborate form of performance art, infused with tricks and rituals that would have seemed laughable back on earth. It was all part of their adaptation. They learned that tortillas were better than more usual breads because they didn't leave behind as many instrument-clogging crumbs, and drinks were always mixed in bags and taken through a straw to avoid spills. Sitting down at their galley table with the aid of those foot restraints on the floor, Expedition Six saw even snack time become as choreographed as a dance recital.

In the afternoons, Pettit liked to eat honey and tiny, addictive loaves of brown Russian bread that he called Barbie bread, because it looked like something she might have pulled out of her Easy-Bake Oven for Ken. But he couldn't just carelessly cram a fistful into his mouth. First, he would open the bundles of bread but leave them tucked away in their package, slipped under a Velcro strap that stretched across the table or pinched in a clip on the side of it. Then he would dig out a tin of honey about the size of a can of cat food from a galley drawer, find the can opener, and keep a set of wet wipes at the ready, just in case some stickiness escaped. He had learned early on to keep the lid attached by a small tab, because that left him one less thing to worry about. He would wipe the can opener and stow it away, and then he would pry open the lid with a set of chopsticks that he had brought up with him. Amazingly enough, he learned that he could stick the tin to the table with just a few drops of water—because of the magic of surface tension, it acted like glue. Suitably prepared, he'd pick the little loaves of bread out of their package and dunk them in the honey, where they would stay until he rescued them one by one with his chopsticks and popped them into his mouth. (Sometimes he would pull out the bread just far enough to coax a long string of honey from the can, careful not to break it, and then let the bread go, watching it fall as slowly as an autumn leaf, caught in the honey's web and the spirit of weightlessness.)

On those days when he felt like whipping up something partic-ularly creative, he'd snip open a small, single-serving packet of peanut butter, squeeze it into the honey, and mix it up with his chop-sticks. Once, filming his routine, he caught himself salivating while stirring the gooey mess. "It doesn't get any better than that," he said with a smile. To keep his peanut-butter-and-honey plague from spreading, he'd leave his chopsticks stuck deep into the concoction, a social faux pas that he apologized for, but one that he figured the Japanese would forgive him.

When he was done, he would fill a garbage bag reserved for dry waste. On those rare occasions when some traces of food were left over, they would be sealed away in a separate bag held closed with a rubber band, to keep the rot (and smell) down to a minimum. And anything even remotely moist, like the wet wipes, were left out to dry, every last drop having become an invaluable commodity, almost as precious as Pettit's chiles. In time, the moisture would evaporate into the air and get caught, cleaned, and distilled by the recycling units. Expedition Six would end up drinking the "wet" from their wet wipes—along with their breath and sweat—sometime later in the week, perhaps boiled up in a nice cup of tea.

. . .

These were the sorts of lessons they learned, the hearts of their new routines. By the time Expedition Six neared a month in orbit, their days on station had taken on the pleasant rhythm of a cross-country drive, an almost reassuring sameness. Morning, afternoon, and night, the temperature remained a perfect constant; the views were reliably spectacular. Bowersox, Budarin, and Pettit had each begun to master the art of living in his brave new world. They had learned secrets and shortcuts, and each had begun to practice his own magic, finding enough ordinary comforts to ground his extraordinary existence. Bowersox had found that his exercise blissed him out like yoga might, especially when he put on his headphones and disappeared in his music; Pettit had come to like eating his breakfast in bed, maybe checking his e-mail between long stretches of looking out the window; Budarin busied himself with learning a little more English each day, usually adjectives that would help him describe his contentedness. In that, he wasn't nearly alone.

Expedition Six had the occasional craving, for showers, for coffee drunk out of a cup, for the touch of their wives, for the sound of rain hitting their windows. They might wake up from a dream with their minds latched on to something as simple as the smell of a glass of red wine, and they wouldn't be able to shake the desire until, like a song stuck on heavy rotation in their heads, it was replaced by a new one. Each of them, then, had his bouts with loneliness and his low moments. But they were rare. These were three joyful men, having struck a fine balance between the epic and the everyday.

. . .

There were only occasional disruptions to their patterns—most of them happy but still enough of a jolt to remind them that, somewhere down there at least, a more real life was still going on.

Christmas, for instance, came nearly five weeks into their time in space. They often lost track of the days (without the changes in seasons or the length of their nights to help wind up their body clocks, sometimes they had trouble knowing what month it was, let alone whether it was a Tuesday or a Wednesday). But calendar landmarks, especially the high holidays, always leaped out. They pulled the men through their mission like the footholds on a rock face.

After long, warm conversations with their families and friends, Expedition Six held a teleconference with the ground. The three of them, surrounded by racks of laptop computers and chattering instruments, donned Santa hats (someone had been thinking ahead) and gathered in front of a camera. Bowersox was in the center, Budarin bobbed up and down to his right, and Pettit floated to his left, along with a small, sparsely decorated Christmas tree, spinning freely.

Bowersox spoke first: "Greetings from the Expedition Six crew aboard the International Space Station," he began. "We may be only 250 miles from home, but it feels like the longest 250 miles in the universe, and those miles seem even longer now during the holidays. So we'd like to take the opportunity to send our best wishes to all on earth during this holiday season."

He handed the radio to Pettit. "Of course, we'd all love to be home with our families for the holiday season," he said, allowing himself to think about his twins opening their presents, if only for a moment. "But working here on space station is all part of exploration: exploration of our solar system, our planet, and ultimately, us as human beings. And this will bring us clo—"

Just then, Pettit's transmission cut out, his deepest thoughts lost to time and space.

Finally Budarin, perhaps mindful of the spotty connection, took hold of the radio, opting to keep his greetings short and not so sweet. (New Year's matters more to most Russians anyway.) He recited a grim-seeming proverb that's been long passed down within the cosmonaut corps. Roughly, it goes, "First, we live in a crib. But we cannot live in a crib forever." He might as well have said, "Merry Christmas, now grow the hell up." Perhaps in its original Russian, it sounded more like poetry and less like an order.

He passed the radio back to Bowersox, who tacked on a more uplifting finish: "We wish our families, friends, and everyone back on earth peace, joy, and goodwill. From the three of us here in space"—and then the rest of them chimed in, shouting in unison—

"HAPPY HOLIDAYS!"

The festivities continued. After hanging a white cloth banner of a homespun Christmas tree, and with some strange Russian techno music providing the soundtrack, Expedition Six decided to try making a cake. With no oven at his disposal, Bowersox retrieved a few Twinkies he had asked the ground to supply. (The ground had obliged, but long before Expedition Six had made it to station; the Twinkies were at least six months old, probably older.) He laid them out in the vague shape of a candy cane on a piece of cardboard, using dabs of the frosting that had also been shipped up for glue. He spread the rest of the frosting across the Twinkies in pink and white bands. Pettit, floating over the finished product with his video camera, zoomed in for a closer look at the creation. After a few seconds of contemplation, he rendered his verdict: "Actually, that's pretty good."

Bowersox, however, humbly admitted that it looked more like

the number 1, which it did. "Our number-one Christmas in space," he said a little sheepishly, and with that, the three of them began digging in with knives, using them like shovels.

"Not bad," Pettit said, wiping icing from the corners of his mouth.

That cake tasted like home.

. . .

Another month passed, the days and nights lost in an endless string of orbits. During one of them, Expedition Six reached the halfway mark of their mission. They were deep enough into their journey to make it harder and harder for them to remember its beginnings. Instead, their minds had become preoccupied with its end—not that they wished for it, or that they even looked forward to it, but for the first time they were mindful of its coming. Now, whenever they stared through their windows, what once had seemed so far away looked as close as it had ever been.

On February 1, 2003, Don Pettit began his day as he always did, by fogging glass with his breath. Below him, in a small corner of Florida, the last of the mist had burned away to reveal a perfect morning, cool but comfortable, with just a few clouds stretched across the opening sky. The bleachers that had been set up about halfway down the Kennedy Space Center's three-mile-long runway caught the sun and began to fill. Reporters, dignitaries, and the husbands, wives, and children of the seven astronauts on board the space shuttle *Columbia* gathered to watch its gliderlike return after a successful sixteen-day mission. In front of the bleachers, a large digital clock ticked down toward the crew's scheduled arrival time. Rick Husband, Willie McCool, Kalpana Chawla, Laurel Clark, Michael Anderson, David Brown, and Ilan Ramon were only minutes away.

Aside from their shared inexperience—Husband, Chawla, and Anderson had each flown into space just once before; the rest were rookies—the crew assembled for STS-107 was as diverse as any NASA had put together.

Husband, the mission's commander, was a forty-five-year-old

air force colonel, a devout Christian, and a graduate of the Test Pilot School at Edwards Air Force Base. The native of Amarillo,
Texas, was also the married father of two.

Sitting next to him in the pilot's seat was McCool, a Naval
Academy graduate (he finished second in his class) and aircraft carrier flier. Born in San Diego, the forty-one-year-old had been raised
in Lubbock, Texas. In his youth, he had been a top long-distance
runner. Along with Husband—along with every member of the crew
except for the bachelor Brown—McCool was married, as well as the
father of three boys.

Behind Husband and McCool sat Chawla and Clark. Chawla,
acting as the shuttle's flight engineer, was an aerospace engineer and
an accomplished pilot, fond of stunt flying. A native of Karnal, India, the forty-one-year-old was a hero to some in her homeland,
where many women couldn't dream of one day reading a book, let
alone strapping in for a flight into space.

Clark, also forty-one, wasn't quite so removed from her Racine,
Wisconsin, home. A navy commander, diver, and physician, she had
become the mission's self-styled documentarian, with plans to film
the entire descent. She hoped the upbeat footage might help her earn
the forgiveness of her eight-year-old son, Iain, who had begged her
not to make the trip. The family had recently survived a small plane
crash unscathed, at least physically. But deeper down, Iain had
scars.

Below deck, where Bowersox, Budarin, and Pettit had been
locked away for their launch, the three remaining crew members
stared at the same rows of storage lockers.

Anderson, the mission's forty-three-year-old payload commander, called Spokane, Washington, home. He was one of the few Afri-
can American astronauts in NASA's pool. He, his wife, and their two
children attended the same church as Husband. Anderson believed
in Heaven.

Brown was the only former circus acrobat on board. The forty-
six-year-old native of Arlington, Virginia, was, like so many of his
crewmates, a product of the navy, an aviator, and a flight surgeon.
Unlike the rest of them, he had gone into space certain that he

would not return. In the weeks and months before launch, he had been plagued with premonitions that his first flight would be his last. He had gone so far as to tell his friends, if not his crewmates, that he would not be coming home.

And last there was Ramon. He was the oldest member of the crew, forty-eight, and he had the most children, four. The son of German and Polish refugees (his mother had survived Auschwitz), Ramon was born in Tel Aviv to a new world. He held a degree in electrical and computer engineering and had been an ace fighter pilot, leading bombing runs into Iraq and Lebanon. He preferred not to talk about them. He was more open about his becoming Israel's first astronaut. His surprise selection—he hadn't even applied for the honor; it was offered to him in a phone call out of the blue—had left him humbled and thankful. "I think I was in the right place at the right time," Ramon said.

He wasn't content simply to hitch a ride, however, the way some foreign astronauts had been (namely the Saudi prince stowed in the bowels of *Discovery* in 1985). He had asked to perform an experiment that meant something to his country and his people. After much debate, it was decided that he would study how the dust picked up by winds across the Sahara affected weather in the Mediterranean.

The dust analysis was one of more than eighty experiments that had been scheduled for the mission, most of which took place in a double-wide Spacehab module dropped into *Columbia*'s cargo bay. They included the usual investigations into bone loss and the physiological effects of weightlessness. There was the latest episode in the seemingly endless survey of zero-gravity protein crystal growth. New technology in space navigation, satellite communication, and thermal control systems was tested. The amount of solar radiation reaching the earth and what was left of the planet's ozone layer were measured. A sample of xenon was carried into space to watch how it behaved in low temperatures. A small zoo was also carefully tended—the proverbial guinea pigs, albeit in the shape of thirteen rats, eight spiders, five silkworms, three carpenter bees, fifteen harvester ants, and a school of fish.

It was a full load, partly because *Columbia*'s flight would be the last devoted exclusively to scientific research. Until the fleet was finally scuttled in a few short years, every other shuttle mission would visit the International Space Station to help finish its assembly. That distinction left STS-107 subject to intense prelaunch criticism; some felt that the flight's $500 million price tag was too high given the expected returns. NASA officials argued otherwise, but their protests began to ring hollow when the mission was repeatedly bumped, from July 2001 to July 2002, until, finally, to January 16, 2003. The discovery of cracked fuel lines and frayed wiring in the notoriously prickly *Columbia* contributed to the delays, but so, too, did two favored missions to the Hubble telescope and the continued treks to station. That's how STS-107 came to lift off nearly two months after Expedition Six had on STS-113. Rick Husband and company had been repeatedly tapped on their shoulders and pointed toward the back of the line.

The crew had tried to make the best of their sometimes torturous dragging out. For more than nine hundred days, Husband, McCool, Chawla, Clark, Anderson, Brown, and Ramon had worked toward their shared goal of finally reaching space. They even found the time to lift themselves at least part of the way there, scaling 13,000 feet to the top of Wyoming's Wind River Peak. They had hoped that the climb would boost their spirits, and it did. On top of that mountain, they were reminded of all that they were waiting for.

• • •

Liftoff had been seemingly flawless, as had been the mission. Over the course of 255 orbits around the earth, the only snag had come when one of the Spacehab's air-conditioning units sprang a leak and, to avoid the risk of condensation building up in the module, the unit was shut down. Given their previous hurdles, the crew wasn't about to complain about a slight spike in temperature.

During their busy time aloft, they had stopped only once, on January 27 at 12:34 p.m., to call up Pettit, Bowersox, and Budarin. ("We're really excited to be able to talk to you guys, one space lab to another big old space lab on that beautiful station of yours,"

Husband said.) Pettit was probably the closest to the shuttle crew; McCool, Clark, and Brown were classmates of his. The rest knew one another only casually. But over the preceding days, they had forged a deeper bond. Six billion people were on the planet. Only ten were in space, and they knew that together, they were virtually alone, united in their isolation. Ramon had promised to hug Bowersox's three children for him after his return. Pettit, looking down at the Black Sea, and McCool, orbiting over Brazil, had been involved in a longer dialogue.

In his waiting for his own mission to get off the ground, Pettit had designed a chessboard (patent pending) made of the soft half of a square of Velcro. He had then cut out white and black pieces from swatches of the sticky half. By e-mail and over their radio, Pettit and McCool had announced their next moves, each prying their pieces from their respective boards and pressing them back into place. The game was made one for the record books by the distance between them. All that was going on, and they could still trade pawns.

During the single conversation between the shuttle and station crews, McCool had been scheduled for sleep. Before lights-out, however, he had asked Husband to relay his move to Pettit, and Husband had obliged: E2 to E4.

Before hitting the sack himself, Pettit moved the piece and stared at his makeshift board, reflecting on McCool's latest play. He went to sleep thinking about his next move.

. . .

Five days and four nights later, when Pettit woke up again to his life's beautiful sameness and fogged over his window, he and the rest of Expedition Six knew in the backs of their minds that *Columbia* was to return to earth, but traveling in a vessel that was eight times faster than a rifle bullet didn't hold the same awe for him as it did for the crowds gathering on the ground. Across the southwestern United States, shuttle watchers switched off their alarms, stepped outside into the chill, and turned their cameras and telescopes to the sky, waiting for a white light to streak across it.

In Florida, the bleachers were now nearly full. A few of the children played behind them. The husbands and wives talked about their plans for welcome-home meals, maybe a drink or two, and some overdue time together on the couch, hearing stories of magic and impossibility.

In *Columbia*'s cockpit, Husband and McCool monitored the instrument panels. The shuttle's descent is automated, its safe return one of the marvels of physics. The friction from the atmosphere conspires to slow it and drop it, bit by bit, toward home. McCool had been looking forward to reentry; he had heard so much about the accompanying fireworks, and now he would finally get a chance to see them with his own eyes.

Northwest of Hawaii, Columbia dropped below 400,000 feet, pushing through the first molecules of the upper atmosphere. The few early sparks didn't impress McCool as much as he had hoped. He felt let down. But as the shuttle continued its descent, the fire outside its windows continued to build.

"It's going pretty good now, Ilan," McCool said, trying his best to describe the view for his friend below decks. "It's really neat, just a bright orange-yellow out over the nose, all around the nose."

In time, the bright orange-yellow turned into a full-blown inferno.

"You definitely wouldn't want to be outside now," Husband said.

"What, like we did before?" Clark joked, distracted for the moment from her filming. She returned her focus to the camera's viewfinder, capturing the smiles of her crewmates while they charted their course over Hawaii, across the last patches of the Pacific Ocean, into the airspace over northern California . . .

Mission Control noticed abnormal readings from four temperature sensors in the shuttle's left wing.

. . . over Nevada, Arizona, New Mexico . . .

Husband called down: "And, uh, Hou—" His transmission was cut off.

Mission Control saw then that more sensors had tripped, indicating a loss of tire pressure in the left landing gear.

Husband tried to talk to the ground again. He had seen the lights go off in front of him: "Roger, uh, buh—"

. . . on into Texas . . .

When, just sixteen minutes before touchdown, all of those shuttle watchers on the ground saw that heartbreaking flare, and that one streak of white light becoming several.

But Mission Control couldn't see what those sky-turned eyes had seen.

They knew only that on liftoff, just eighty-one seconds into *Columbia*'s flight, a chunk of the external tank's insulating foam had broken off, striking the underside of the left wing. Over the course of the crew's sixteen days in orbit, film of the foam strike was watched again and again by engineers on the ground, just to make sure that no serious damage had been done. They decided that it wasn't cause for concern. Wayward foam had struck every shuttle during launch. Always, it had bounced away harmlessly, like a bug off a windshield.

This time, however, it had not been harmless. The foam had punched a ten-inch hole into something called RCC panel 8, one of the black, heat-resistant, reinforced carbon-carbon panels that cover the shuttle's nose and the leading edges of its wings. The same superheated plasma that had enraptured *Columbia*'s crew poured through that hole like mercury, burning away the sensors that first raced hearts at Mission Control. While the shuttle continued its journey home at eighteen times the speed of sound, thirty-seven miles up, that plasma melted the wing's aluminum skin from the inside out. Without it, *Columbia* first began to shake, and then to tumble, and finally it broke apart.

"*Columbia*, Houston, comm check . . ."

There was no reply.

There was only quiet. In their desperation, technicians willed Husband's voice to crackle across the radio, for streams of data to begin pouring out of the heavens, for a blip to appear on a screen so that everybody could breathe again.

Outside, in the bleachers at the start of this seemingly perfect day, the adults waited for the two sonic booms that would signal *Columbia*'s arrival, one after the other, two minutes before touchdown. The countdown clock ticked past that deadline, and still there was no sound. In a growing silence that was broken only by the sounds of the children playing, the crowd watched the clock continue its countdown, second by second. Before they could watch the clock reach zero, the families were taken by their hands and loaded into vans and smothered in hugs.

In a family video conference during *Columbia*'s final flight, Laurel Clark's doom-fearing son, Iain, had asked the question that even those who knew intimately the answer now found themselves asking:

"Why did you go?"

. . .

Still warm in his sleeping bag, Pettit found himself thinking about coffee. As it had for the Russians before him, it was becoming a concern. He didn't have anything like one hundred pouches anymore. But looking out through his window at another orbital sunrise, he decided to hell with it: this was the sort of morning that coffee was made for. He put on his glasses, pulled himself out of his sleeping bag, pushed his way out of his private quarters, and found his center of gravity. With it, he propelled himself in clean, practiced movements, like a swimmer who's found his stroke, past a sleeping Bowersox, toward the far end of station. There, Budarin remained zipped away. Pettit opened the metal drawer that held his fixes and took out a pouch, a silver bag with powder packed hard into the bottom of it. He filled it with hot water and began hunting for a straw. He found one, as well as a little Russian *tvorog* to eat, and headed back for his sleeping bag. Pettit tucked himself in, savoring another breakfast in bed, and turned his mind toward the day ahead. It was a lazy Saturday. He'd finish his coffee, check his e-mail, and then spend the rest of his weekend cleaning house— scrubbing fingerprints from windows, wiping down handrails with antiseptic solution, even mopping up the occasional coffee splatter.

But there wasn't any hurry. Although he had sometimes felt like it was flying, he knew that time was the single thing he wasn't really running out of. He was still weeks away from home.

Pettit took another sip and watched the sun rise for a second time. The rush of it still drew him to the window, the sun coming and going every forty-five minutes, good for sixteen dawns and dusks a day. Next he looked down at the vapor trails that were folding on top of the United States like a quilt, the way they always did, one by one by one, New York to Los Angeles, Boston to San Francisco. They had become his way of catching a glimpse of home even when it was shrouded in storms. But on this day, the horizon was clear and the sun was bright, so bright that he didn't notice the finger of white smoke spreading out over Texas.

He finished his coffee. He got up and began puttering, checking his watch every so often to make sure he didn't miss the ground conference scheduled for that afternoon. At about two o'clock, Greenwich mean time—sailors' time, the official time zone of station—there was a conversation planned with Houston to draw up next week's activities. Usually the voice coming out of the radio told the crew what they already knew, and they floated about, keeping their ears half open for news or drama. This time was different. This time, the voice told Expedition Six to stand by.

There are two main rooms at Mission Control, next door to each other, almost identical in design but now cast under different shadows. In the first, from which the shuttle is commanded, the silence had turned hopeless, watering eyes turned toward the end of *Columbia*'s last orbit, a line left incomplete, frozen on the giant screen at the front of the room. In the second, from which the space station is tracked, and where numb technicians sat behind consoles labeled ODIN, OSO, ECLSS, ROBO, and a dozen other things, a heated debate was unfolding. No one was sure how to tell Expedition Six that *Columbia*, the shuttle that Bowersox had twice piloted, had just disappeared in the thin blue-green envelope beneath them. No one was sure how to tell them that seven friends were probably gone, too.

Jefferson Howell, a retired marine lieutenant general and the plainspoken director of the Johnson Space Center, ended the debate when he sat down at the radio. He considered his words for only a moment before he pressed a button that would bounce his voice off a satellite and into the space station's tinny air.

"I have some bad news," Howell began, and because it was Howell who was delivering it, Pettit and Bowersox knew exactly how bad before he got the rest of it out: "We've lost the vehicle."

Nine words. That was all. Everything else was left unspoken, and in the quiet, the blanks were left for each of them to fill on his own. In the way the parents of missing children hang on to the smallest chance that their loved ones are just lost, not lost for good, Pettit and Bowersox wondered whether any of *Columbia*'s evacuation systems had triggered, and whether any of their friends were floating down to a cloudless earth under parachutes.

They held on to that faint hope until a battered helmet was found on the grass later in the afternoon. The flight data recorder was also found. So was Laurel Clark's videotape.

Each grim discovery was reported to Expedition Six. Each pushed aside their faint hope to make room for more sadness.

Pettit folded away his chessboard, finally knowing that the game would remain forever unfinished.

The sadness settled itself in.

. . .

Every so often, Bowersox, Budarin, and Pettit had been allowed to use the Internet phone on closed channels, with the tape recorders turned off. This was one of those times. Before, when no one was home—when it was time for her to get the groceries or for the kids to go to soccer practice—they had left messages on the machines: "Hey honey, it's me, in space." Sometimes those messages were saved and listened to in the still of the night, again and again. After the conversations of that afternoon, those messages would almost always be saved, because she never knew when they might become all she had left.

In that way, their families had finally caught up. In their time away, the men of Expedition Six had learned already that the everyday interactions of life on earth—the messages left on machines, but also the smiles and waves from school buses and the notes left on fridges and pillows—were the things worth carrying with them up into space. They had made room for them in their memory's permanent collection, just as they had learned to forget about the trivialities that they had once kept too close. And they had come to understand the true order of things, because they had learned how the universe works. Some astronauts become the first men to walk on the moon, and others burn to death sitting on the launchpad or seventy-three seconds after leaving it, or sixteen minutes from returning to earth.

Now they knew, too, that they were no longer weeks away from home. The next shuttle up was to take them down. But they remembered *Challenger*, lost nearly twenty years ago, and they knew that their ride wasn't coming anytime soon. They were suddenly much farther gone, although they weren't really sure how far, because just like that, the miles were made more meaningless than ever before. It was distance without measure. There were instants when Dallas was farther away from Houston than they were. But what mattered now, what separated them from home, was time. Suddenly they were locked in a souped-up Airstream, trapped on the other side of that single pane of glass.

They told Mission Control that they were all right, that they had trained a lifetime for this, that they could hold on to their memories for another year, maybe longer. Part of them might even have believed it.

But in the coming dark days, after *Columbia*'s memorial service was piped in from the ground—after they had heard President Bush say, "Their mission was almost complete, and we lost them so close to home"—and after they'd rung the ship's bell in Destiny, seven times for seven astronauts, they couldn't help thinking that their friends hadn't been so close to home after all.

3 THE GEOGRAPHY OF ISOLATION

But this is how close they were:

The men and women of *Columbia* returned to earth the way they had left it, accompanied by sonic booms and vapor trails. The first debris reports were called in from the small town of Lufkin, Texas, only a three-hour drive east of the crew's homes in Houston. Heat-scarred wreckage landed in forests, fishing holes, and parking lots. Smoke rose from front lawns. A piece crashed through the ceiling in a dentist's office. An engine fragment weighing six-hundred pounds hit the sixth fairway at a golf course with enough velocity to break through the water table and create a small pond.

Frank Coday was sitting in his mobile home in Hemphill, Texas, smoking a cigarette, when he heard strange noises outside and pictures began falling off the walls. On the uneasy hunt for answers, he went outside, casting his eyes up at that perfect blue sky. He decided to climb into his truck and make the short drive across to his brother's house, down their shared country road. On the way, he saw an unmistakable, lifeless shape on the gravel in front of him. He steered around it and picked up the pace.

His brother, Roger, had heard the same noises and was waiting for him. Frank told him that he'd seen something on the road, something that he didn't want to look at alone. The two men drove back and got out of their truck. There, at their feet, were the remains of an astronaut.

There were more cradled in the oak trees around them. And more still in the corner of a nearby pasture.

Jim Wetherbee, who had volunteered to head the recovery operation, was already on his way from Houston. Soon he would pay a visit to the Coday brothers. Over the next five days, Wetherbee's team found the remains of all seven astronauts. With each new nightmare report, police tape went up. Reverends issued last rites. The remains were carried away in hearses driven by local funeral directors and later flown to Delaware for identification. Finally, they were returned to their families.

But in some way, those astronauts had arrived home long before, in the instant they had found rest in the East Texas hills. They had traveled millions of miles and somehow, as if by fate, they had wrapped their way around the world and into the arms of these wide-open spaces, the big sky country that astronauts usually come from and to which they inevitably return: to places like Lufkin, Hemphill, and Littlefield, all of those prideful small towns that the wind has scattered across the landscape like seeds. From space, when darkness spreads across them, they light up one by one, looking so much like stars laid against the night.

· · ·

Don Pettit was born and raised in Silverton, Oregon, another one of those seeds, another one of those stars. There are farmers there and loggers and Pettit's father, Virgil, was their doctor. He did a little bit of everything, from pulling out gallbladders that looked like bags of marbles to delivering babies, all for what his patients could afford on that particular afternoon. He would receive letters from the American Medical Association suggesting strongly that he raise his rates, but Dr. Pettit always believed that fast pennies were better than slow dollars, and besides, the way he ran things made him feel good. Don, the youngest of his three sons, would go down to the office and watch him work; he took the theory of his father's practice to heart. He also took to pulling out the old, brown skeleton that sat jumbled in a bottom drawer and trying to piece it back together. It's hard to know, exactly, but maybe that's where Pettit the Younger picked up his tic for making parts into their intended whole. Later, one of his brothers gave him a two-speed transmission for his birthday, and Pettit still remembers it as the best present he's ever received. He spent entire afternoons taking it apart and putting it together until he could knit it through by touch.

Transmissions gave way to windup clocks, windup clocks to the logging equipment that he repaired during his summers off from school.

Not surprisingly, when it came time for Pettit to head to university and choose a course of study, he picked something that would give him puzzles to solve: chemical engineering at Oregon State University in nearby Corvallis. There he pulled apart problems instead of machines, honing his natural bent toward analytical thinking, but he was also being encouraged to tap the unconventional streak that ran through him, mostly by a professor named Dr. Octave Levenspiel. When devising a conundrum about gas molecules and entropy, Dr. Levenspiel would find a way to make it about canaries in a cage. Teaching the fine art of approximation, he would ask Pettit how many barbers plied their trade in New York City. Like a magician training an apprentice, he taught Pettit tricks. He stretched him, too, and groomed him for big dreaming—engineers had picked up from God in building the world. Suddenly all of the things that Pettit was obsessed with as a kid—airplanes, electric trains, and rockets, especially—came back into play, part of the larger equation. Nothing was out of his reach, not even space.

He had spent countless childhood nights staring through his brother's cheap telescope, trying to pick out planets that looked more like fuzzy footballs. And he had listened again and again to John Glenn describing how earth looked from orbit, the astronaut's voice scratching through a free floppy record that came with Pettit's new pair of Red Ball Jets.

A throw-in with his sneakers made him want desperately to look over Glenn's shoulder and enjoy the same view. With the help of Dr. Levenspiel, childhood fantasy seemed that much closer to coming true. The universe started to look like that old two-speed transmission, like just another machine for him to find his way through.

· · ·

Like Pettit, Ken Bowersox came of age under orbits. When he was eight years old, he sat in the front seat of his father's car, the pair of them tooling around their hometown of Bedford, Indiana, glued to the radio broadcast of John Glenn's trip around the earth. Along with a million other Midwestern kids, Bowersox decided over those fantastic few minutes that he wanted to become an astronaut. Unlike most of the rest of them, he never changed his mind. A few years later, in junior high school, his class was given boilerplate career handouts that outlined the best way to pursue every job then invented. Bowersox immediately grabbed the sheet that pointed toward space and committed the directions to memory: he needed to attend a military academy or a college with a strong foundation in engineering, complete exemplary military service, and finally seal the deal with time served as a test pilot. In that moment, his life's course was charted as precisely as if he were following points on a map.

His first stop was the United States Naval Academy in Annapolis, Maryland. At first glance, with its grassy courtyards, mature trees, and collection of redbrick colonial buildings, the campus looks like any other upmarket Eastern school. But deeper inspection reveals all of the small things that separate it from Harvard or Princeton. Through the blankets of cold rain that so often lash the Chesapeake coast, students' white caps stand out against dark skies and uniform coats; a guardhouse watches over the front gate; cannons and anchors are scattered about the yard; a massive bronze bust of Tecumseh takes another beating from the elements. Founded in 1845, it is the navy's answer to West Point, where smart kids with short haircuts find the resolve to captain ships. Their rivals in the air force might dismissively refer to the school as Canoe U., but not surprisingly, the academy has proved this country's most fertile ground for harvesting astronauts. To date, more than fifty have been born here as if stamped from a factory mold—which, in some ways, they have been.

Bowersox, like the students before and after him, began his four-year Officer Training Program on Induction Day, the start of the daunting "Plebe Summer," a seven-week boot camp for incoming recruits. From their first hours on campus, midshipmen are taught how to wear their crisp new uniforms and how to salute. They also take their oath of office, promising to abide by the academy's honor code: "Midshipmen are persons of integrity," it begins. "They stand for that which is right." The rest of their summer days begin at dawn with a rigorous program of exercise and drills; lessons in seamanship, navigation, and boat handling; as well as training in small arms. (It is not the usual orientation week by any stretch.)

The rest of their undergraduate education is built on a similarly strict regimen dedicated to making them fit for command. The 4,000 students are divided into thirty companies; for the next four years, they eat, sleep, study, drill, parade, and compete as a well-disciplined unit. All hands are up at half past six. Each day is filled with six hours of class. Each student takes a core curriculum in engineering, science, mathematics, humanities, and social science. In between, they march to meals en masse. Afternoon athletics are mandatory. (It's not unusual to see a group of uniformed midshipmen running around with logs or boats hoisted over their shoulders.) Twice a week, there are yard drills and parades. Without fail, lights are out at midnight.

It's a demanding routine that Bowersox slipped into easily. He liked the discipline of it, the feeling of falling into bed exhausted, having squeezed the most he could have out of a day. He thrived academically, became addicted to long, solitary evening jogs, and demonstrated a quiet leadership. On those rare occasions when he needed a boost, he stopped by the campus museum to look at the portraits of those academy graduates who had gone on to become astronauts: Alan Shepard Jr. (1945); Wally Schirra (1946); Jim Lovell, Tom Stafford, and Donn Eisele (1952); William A. Anders (1955); Charles Duke Jr. (1957); and dozens more since. In the display cases nearby, there remains the small blue Naval Academy flag that Shepard took on board *Apollo 14*, and the sheet of brown paper inked with gold block letters (BEAT ARMY!), carried on *Geminis VI and VII*. These artifacts gave him hope, as if space wasn't all that far away.

He was also picked up every time he walked in and out of Bancroft Hall, the giant, cathedral-like dormitory that he called home. There, lost among the stone columns and the vaulted ceilings and with the last of the day's light shining through stained glass, he listened to the echoes of choirs singing before supper. He took in the vast mural of the Battle of Santa Cruz, bombs splashing in the water. And there in Memorial Hall, reflected in its polished wood floor, he saw the battle ensign flown by Commodore Oliver Hazard Perry at the Battle of Lake Erie, on September 10, 1813. In tall white letters, the commodore had stitched onto a big blue flag the dying words of James Lawrence, captain of the USS *Chesapeake*: DON'T GIVE UP THE SHIP.

By the time Ken Bowersox threw his white cap into the air at his graduation, class of 1978, those five words had been stamped onto his heart like a tattoo.

. . .

After earning his master's degree in mechanical engineering from a more conventional school—New York City's Columbia University—in 1979, Bowersox headed to the Naval Air Station in Pensacola, Florida, and over the course of two long, demanding years, he learned how to fly. Almost immediately, he was posted to Attack Squadron 22 on board the USS *Enterprise*, the world's first nuclear-powered aircraft carrier and still, more than four decades after its christening, the world's largest. During its historic service, it has blockaded Cuba, patrolled the Mediterranean, anchored off Vietnam (ultimately playing an integral part in the evacuation of Saigon), charted course through the Suez Canal, and made a historic 30,565-mile trip around the world without refueling or replenishment. It is a monster, and in the cockpit of an A-7E, on the deck of the navy's most storied ship, cruising the Western Pacific, Bowersox found himself exactly where he had wanted to be.

Despite sharing his post with 4,600 crewmates, he soon learned that life on a ship was an exercise in loneliness. Waving goodbye to families gathered on shore, the sailors and pilots lined up on deck routinely headed for voyages six months long. On those rare occasions when they visited land, it was sometimes hostile, and they were almost always strangers to it. And for as long as they were at sea, the satellite phone was off-limits except for emergencies, and mail from home was usually a month behind, an information lag that gave aches. Like marooned castaways, the sailors were separated from their former lives and loved ones by thousands of miles of bottomless ocean. Together, they became hermits and monks.

The pilots among them became projectiles, too. Landing on an aircraft carrier, even one as large as the *Enterprise*, was like getting fired out of a cannon into a very small net. The runway was short, narrow, and moving, probably forward at speed, and probably with tremors and rolls thrown in for good measure. Each time he came in for touchdown, Bowersox would turn in toward home, listen for corrections to his flight path, make sure his wheels were dropped, and deploy his tail hook—the flimsy-seeming catch that, if everything went the way it was supposed to, would grab hold of the arresting gear, four steel cables lashed across the decks. He would hit the deck with enough force to blow out smoke, and just in case his tail hook missed the mark—just in case his A-7E became a "bolter," in navy parlance—he would land with his engines at full power, giving him just enough juice to take off again if he went unstopped and reached the front of the ship. Otherwise, he would fall into the deep end of the ocean, and all of his dreams might drown with him.

More than three hundred times, Bowersox slammed down on the deck of the *Enterprise*. And more than three hundred times, he came to a jarring halt. He had it down so cold, it began to feel as though he was pulling into his driveway, another night safe at home.

Eventually, however, the Midwestern boy in him decided that it was high time to get his land legs back. Having become all too used to staking his life on a length of steel cable, he chose to fly as far as he could from life at sea.

. . .

The desert draws out the peculiars, mostly because it takes a special breed to live in a place that doesn't want the company. That's why we bring our worst prisoners here, stashing them behind high fences and miles of suicide country, and it's why we put signs on the road-side warning late-night drivers not to pick up hitchhikers, especially those still wearing jumpsuits and foot restraints. But it's also why so many more come here by choice, even if it's sometimes a desperate one. They like that darkness falls harder here than anywhere else in the world, the last drops of moonlight swallowed up by the sand, their headlight beams disappearing into the black. They like the emptiness. They like the echoes. They like that their radios some-times go out. They like the mystery of what lurks around the next bend. They like the ghost towns, this landscape of sagging rooflines and boarded-up gas stations. And with it, they like the good feeling that comes with surviving in a place that has sent countless others packing. It's as if they like being surrounded by the starkness of fail-ure, because that makes even their smallest successes feel that much more like triumph.

Perhaps that's why, too, we've always made this place our prov-ing grounds, blowing the almighty crap out of sound barriers and land-speed records and chunks of New Mexico. Of course, there are plenty of other, equally good reasons for exorcising those demons out here. The chances of hitting a house with a mistake are slim. There aren't too many spying eyes. And there are the great salt flats that stretch far enough into the horizon to compensate for the biggest touchdown miss. But even those salt flats look small next to that larger truth: when engaging in a business in which things often don't work out—like flirting with the laws of thermodynamics—it never hurts to stand shoulder to shoulder with a long line of hard-luck cases. Because then, even at your loneliest, you're never really alone. That's why the Southwest will forever remain a pioneer's ter-ritory, and that's why it's crawling with astronauts and the men who might become them.

It doesn't hurt that in the waiting until they get to touch space, space will touch them. Living in the desert is like a test drive for life in orbit. Here, like nowhere else on earth, the line that separates up there from down here is blurred, a vanishing point lost in all of that highway shimmer.

East of Flagstaff, Arizona, down a long, straight road that forks off I-40, just a few miles past the Mobil filling station and the Meteor Crater RV Park, there's a spot that marked first the death, and then the birth, of all that's happened since. Over a rocky crest that rises out of the featureless plain, a hole suddenly opens up, 550 feet deep and more than 4,000 feet across. (If it were a football arena, more than two million spectators could sit on its slopes and take in twenty games at once.) This particular hole was made 50,000 years ago by a massive iron-nickel meteor, which struck the desert hard enough to turn graphite into diamonds. Out of that impact was also born a boundless fascination, first recorded by one of Custer's scouts, a man named Franklin, in 1871, and later shared by curious sheepherders, ranchers, miners, geologists, and, ultimately, astronauts. The Apollo classes came here to test-run spacesuits and buggies, because it was only a small leap between the crater's landscape and the moon's. From the edge of that hole, the universe feels almost uncomfortably intimate, the way it did that morning when *Columbia* and its crew returned to earth.

The reality is, it's part of the natural order of things for objects to fall out of the sky. Every time we launch into space, we're defying the universe's most basic mechanics. Meteor Crater is a reminder of that. So are the myriad strange bits of asteroid lore hidden in the pages of old newspapers, like the story of that evening in 1954 when Mrs. Elizabeth Hodges of Sylacauga, Alabama, found Heaven in her lap. (The Southern lady was in repose on her couch when a meteorite weighing almost nine pounds crashed through her living-room ceiling, bounced off the top of her wireless, and settled in the crook of her matronly hips.) Or, more recently, there was an afternoon in 1992, when a twenty-seven-pound intergalactic hailstone had eyes for a red Chevy Malibu in Peekskill, New York.

Every time something like that happens, a string is tied between us and the stars. When something the magnitude of the meteor that tore a hole into Arizona comes down, that string is more like a corridor, as if a portal has been opened, or a beam of light has been left to track across the night sky the way the glow from Las Vegas banks against clouds, luring gamblers from hundreds of miles around. Whatever it is, it's a magnet, helping the desert draw out not just the peculiars but the downright alien.

There was the reputed 1947 spacecraft crash and Martian capture outside Roswell, New Mexico, which now boasts the UFO Museum and Research Center and the UFO Hall of Fame, as well as an impressive collection of souvenirs: little green men, both inflatable and stuffed, and alien-themed T-shirts, shot glasses, antenna balls, bibs, key chains, bumper stickers, ties, caps, mugs, welcome mats, bandanas, and pots of German chocolate topping. There is Area 51, a long way off Nevada's Extraterrestrial Highway, defined by a lonely black mailbox and signs announcing that deadly force is authorized in the elimination of trespassers. And there is the windswept little town of Rachel, Nevada, and its A'le'Inn, marked by the roadside presence of a flying saucer hanging from an old Chevy crane, and with its smoke-stained walls lined with photographic evidence of unearthly visitations in Chandler, Arizona, and Pacific Palisades, California, and Kanarraville, Utah. (Rachel's ominous population sign—HUMANS 98, ALIENS ?—also invites a second look.) Roll in the Southwest's collection of observatories, air fields, government labs, and plain old black magic, and it's no wonder that Don Pettit and Ken Bowersox were finally drawn here, too. It's pit-stop city for star chasers.

· · ·

Pettit found his oasis at the University of Arizona in Tucson, a low-slung city that feels in some ways like a Bedouin camp, surrounded by red rocks and stiff-armed cacti. He arrived on campus in the fall of 1978, coming up palm-lined University Boulevard, turning left at the fountain in front of the redbrick building they call Old Main, and disappearing into the sprawl of buildings that makes up the College of Engineering and Mines. They are mostly three- and four-story blocks, clean and efficient, softened by cypress trees, gravel paths, and low stone walls. On Second Street, he found his home in the basement of the more imposing Harshbarger Building, where the chemical and environmental engineers hung out. Outside, it hums like a power plant and off-gasses like a nuclear reactor, a sound and a smell that matches its hard, industrial appearance; inside, it's a bit like ducking through a submarine, with open metal stairs connecting floors lined with machinery banked against cinderblock walls.

Students here compete in events like the International Micro Aerial Vehicle Meeting, build submarines to track the migration patterns of the giant Pacific octopus, and research how anaerobic bacteria might one day help us rid the world of toxic waste. The egghead faculty, meanwhile, puts out papers with titles like "A new stereo-analytical method for determination of removal blocks in discontinuous rock masses" or "Multimode decomposition of spatially growing perturbations in a two-dimensional boundary layer." Pettit arrived as the first Ph.D. student for Dr. Tom Peterson, an untenured assistant professor fresh out of graduate school himself. Not knowing exactly how to harness Pettit's wide-ranging interests and energies—although he seemed on his way to becoming an optics jock more than anything else—Peterson set his new charge loose, as Dr. Levenspiel had in Oregon, letting him roam as far as his imagination might take him.

It didn't take long for Pettit to distinguish himself from his fellow students, not in the least because, on the door of his cramped office, he pinned up a photograph he'd taken of his own chromosomes. He brewed his own beer and fermented his own wine, and he collected old electronics. There was also the night out at a bar when, asked why he wasn't eating the chips and salsa that everyone else at the table was demolishing, Pettit borrowed a lighter and lit a chip on fire, holding it up between his fingers and watching the hydrocarbons burn like a candle. "This is why I'm not eating them," he said before blowing the chip out.

But more defining than his admittedly considerable collection of quirks was the almost artful way of thinking he had developed, one that separated his sizable brain from the purely linear, logical minds sported by most engineers. He still had that hyperanalytical circuitry in him, too, but now he also boasted a more creative, theoretical, open-range spirit. He could find new solutions to old problems that most engineers would never have found, because they were lying in wait on a side road that only Pettit was willing to go down.

His crowning achievement was an instrument he patented (with Dr. Peterson's encouragement) called COPS, the coherent optical particle spectrometer. By splitting a laser beam in two, recombining each half, and measuring the shift in phase between them, he had learned how to detect submicron particles floating in the air—tiny aerosols with a diameter that would make a human hair look like heavy rope. Years later, IBM and other companies came out with machines using Pettit's technology; they are still used by scientists and engineers today.

The only hitch came when, frustrated by what research equipment was then available, Pettit set up a glassblowing shop in a closet in the basement of the Harshbarger Building, where he started making custom instruments for his analytical chemistry work. In time, he became accomplished at taking bundles of glass rods and tubes and turning them into distilling beakers or, when he was feeling playful, even ships in bottles. But one afternoon, when he needed to spark a super-hot hydrogen flame to melt some quartz tubing, the tank's regulator hose caught fire, and alarms went off, and the building was evacuated, leaving everybody standing on the grass in the sunshine, looking over at Pettit, wondering whether he had finally managed to burn the place down.

He had not. But almost two decades later, the walls of that closet betray scorch marks, and there are buckets of leftover glass waiting for the next Pettit to come along and turn them into whatever vision he wakes up with that morning.

Dr. Peterson—having become the head of the department—looked at those buckets, shook his head, and said with a smile: "If I could figure out how Don's brain works, I'd patent that, too."

. . .

Bowersox, meanwhile, pursued a different kind of education. His first desert hole-up was the Test Pilot School at Edwards Air Force

Base in outback California, just as that junior-high career sheet had ordered.

Past the two-man guardhouse and through the gate, Edwards looms so big it's almost hard to take in: just beyond the shoulder of a nowhere highway lies a virtual city, dedicated since World War II to the speed trials of men and flying machines. Dominating its center is Rogers Dry Lake, the world's largest, sixty-five square miles of cracked yellow flats that are washed clean by rain and blown glass-smooth each winter, making for nature's perfect launchpad. More than 240 aviation records have been set in the wide-open airspace above it, lending a hardscrabble spirit to the place. These skies have witnessed the first tests of American turbojet airplanes, like the XP-59A in 1942; Chuck Yeager's sonic boom in the Bell X-1 on October 14, 1947; and, in the years after, the deaths of dozens of maniac pilots trying to go faster still. In 1948 alone, thirteen pilots were killed here (including Glen Edwards, the base's namesake), mostly in fireballs. This otherwise featureless desert has been scarred by more than its share of "class-A mishaps," technobabble for smoking holes in the ground.

The early 1950s saw more happily eventful times—and, famously, a lot of smoking, drinking, sex, and car wrecks. Scott Crossfield became the first man to fly at twice the speed of sound. Altitude and speed records were broken and broken again until 1961, when pilots were finally forced to wear pressure suits and oxygen masks in the hypersonic X-15. It reached six times the speed of sound; it also, in the hands of Joe Engle, reached space. He was the first of eight Edwards cowboys to earn his astronaut's wings without ever strapping into a rocket. Those eight men set a new standard, establishing that never again would the pilots stationed here pay mind to gravity. Even today—above the great hangars with rust-colored trim, the gun butts, the dozens of helicopters and planes parked in neat rows, like crops—test crews will purposely put their F-15s into tailslides, flying hard (and loud) enough to make sure that framed pictures never stay level in the houses below. Awesome planes named the Blackbird, the Nighthawk, the Stealth, and the Raptor have been born here, and, in the Test Pilot School, so have the men and women who will fly them. They will learn to push these new planes as fast and as high as they can, and they will sometimes die in them, lost in the building of preflight checklists and the future's routine. In a lot of ways, then, this place is the same as it ever was.

Two classes of student pilots are pushed through each year, a select group of fifty of the best fliers the military has to offer. As always, each class comes up with a slogan steeped in manly bravado ("To Oblivion and Beyond," "Dirty Deeds Done Dirt Cheap," or "No Permanent Damage") and plays an elaborate prank on the other class, usually involving some form of explosive demolition. And, as always, each of the pilots walking through these halls wears the green flight suit like an Armani, with a sort of straight-backed pride.

But somewhere along the way, a different breed of pilot began springing out of Edwards. Over time, there would come to be less smoking, less drinking, less sex, and more sensible family sedans parked out front, replacing the cherry-red convertibles and motorcycles. Unlike Yeager's generation, most of the new breed—Bowersox included—have gone to graduate school, receiving advanced degrees in science or engineering. Most of them, like Bowersox, boast perfect flying records and stellar efficiency reports. Most of them, like Bowersox, take to heart the school's official motto, *Scientia est Virtus*, Knowledge Is Power. And because of that, most of them, like Bowersox, will spend as much time in the classroom—learning aerodynamics, logic, propulsion physics, avionics, and systemology—as they will in the cockpit.

They still learn to fly between thirty and fifty different aircraft during their twelve months here (it didn't hurt that Bowersox was short enough to fit into even the tightest seat); they still explore the outer limits of our most incredible machines; they still develop a respect for wind shear and wake turbulence that borders on the mystical. And yet, there is less swagger in them than in the great, lead-bellied ghosts of the past. They no longer fight their machines or seek to defy them. Rather, they have somehow *become* them, another part that's been tooled into place. The romance has been waived for reason, and at the end of the day, today's students are far more likely to retire to the in-school lounge with their textbooks and a cup of coffee than to some dusty desert bar with black-and-white photographs of dead pilots on the wall.

Instead, they study under pictures of dead astronauts, including a smiling Ellison Onizuka, a fellow graduate killed on *Challenger*.

The newest portrait hanging on the wall by the door is the painting of Rick Husband, class of 1987, just lost on *Columbia*. He had breezed through the Test Pilot School only two years after Bowersox, and yet now he is gone, reduced to art, the victim of one more class-A mishap.

The portrait is a fresh reminder that in a lot of ways—the most important ways—this place really is the same as it ever was.

. . .

Upon his graduation in 1985, Bowersox was immediately assigned to the Naval Air Warfare Center at nearby China Lake. It's a hard-seeming town; it's not the side of California that is captured on postcards. Its heart is the intersection of China Lake and Ridgecrest Boulevards, which sits between two massive, mountainous tracts of land littered with the wreckage of planes. They have been shot down by missiles, new and improved, the latest in Sidewinders, Sparrows, and Stingers. Beyond these desert junkyards, there are pretty peaks in the near distance and patches of raw scrub beauty, but when the light's unflattering and the sky closes up, China Lake can look as if a couple of warheads overshot their intended targets and landed in the center of town. On the left, there's E Charro Avitia Mexican Food, which used to be El Charro Avitia Mexican Food until the "l" fell off the sign out front. Howard's Mini Market has been boarded up and left to despair. The Desert Empire Fairgrounds are dusty and vacant, but China Lake Bail Bonds enjoys considerable walk-in traffic, as does the 40-X Gun Shop and Guns 4 Us (over 300 guns in stock!). A big night out might start at the Golden Ox Charbroil, pass through the Ridgecrest Cinemas—one of the original multiplexes, complete with reclining plastic seats—and end at the barn of a store with the neon sign that cuts to the chase like 90 proof: LIQUOR is all that it reads, red lights on green.

Speeding through town on the way to the coast will yield only a partial view of China Lake, however. (You'd miss the sprawling, fenced-in base for starters, a self-contained Pleasantville, clean and tidy as a ball field.) A look in the phone book is more truthfully revealing of the soul of the place. Under "Taverns" in the Kern County East Yellow Pages, there are only eight entries. Under "Churches," there are twenty-four *denominations* listed—Assemblies of God, Baptist (further subdivided into American Baptist, American Baptist Association, Fundamental, Independent, and Southern Baptist Convention), Bible, Catholic, Roman Catholic, Charismatic, Christian, Christian Science, Church of Christ, Church of Jesus Christ of Latter-day Saints, Episcopal, Fellowship of Christian Assemblies, Foursquare Gospel, Jehovah's Witnesses, Lutheran, Methodist, Nazarene, Non-Denominational, Pentecostal, United Pentecostal, Presbyterian, Seventh-day Adventist, United Church of Christ, and United Methodist—good for eighty-nine houses of worship. That's because for test pilots, touching the face of God is a full-time gig.

When Bowersox arrived here, he promptly began shooting down planes flown by a civilian pilot named Dick Wright. Fortunately for Wright, he wasn't sitting in the planes at the time. He flew specially equipped F-86s and F-4 Phantoms from the ground, like full-scale radio-controlled models. Most of the time, Bowersox would chase these doomed planes and watch unarmed missiles punch holes through their wings, maybe spilling a few gallons of hydraulic fluid. But every now and then, just often enough to keep things interesting, Bowersox would drop in behind one of Wright's drones, lock it in his sites, and launch the latest, greatest warhead straight up its ass. Bits of burning fuselage and engine parts would fall to the desert floor, and Bowersox would follow them down, turning in for a picture-perfect landing, taking off his helmet, and telling the weapons technicians waiting for his good word what he thought of their new toys. He loved the job, loved the excitement and the adrenaline of it, the speed and the power, but most of all he loved it because he remained, more than ever, and more than even most of his fellow test pilots, given to fly.

During those rare times when he wasn't flying for work, he went flying for pleasure. Although a test pilot's salary isn't the sort of wage that lends itself to luxuries, Bowersox, still a bachelor and living on the base, poured every spare dollar into a decrepit twin-engine beater that he'd gun toward the horizon, riding low like a crop duster. His young stud colleagues couldn't see the reason for it; after bolting across the sky in jacked-up F-18s, they saw it like so much of a comedown. Only older fliers like Wright saw the pleasure in it, saw even the God in it. In flights like that, gliding alone on a route run by heart, he saw a kind of beautiful solitude, something meditative in the engine noise and instruments, like riding a motor-cycle through sunflower fields. In his plane, Bowersox was free, untethered, far removed from his roughneck desert home, but even it looked better from altitude. He could see the other side of the mountains and the desert changing color when winter came. He was one of the few men who had seen how China Lake could make for a decent postcard.

On his bravest days, he would push his flying heap higher still, sometimes high enough to catch a glimpse of the curvature of the earth. For Bowersox, moments like those were holier than he could have ever found in any one of those eighty-nine churches, his eyes still catching the last of the sun even while it got dark below, squeezing out just a few more minutes until the lights came on at the airfield and called him home. Flight after flight, he never tired of the view or the mechanics of soaring. He never tired of the solitude either, the feeling that he could just as soon keep flying over the rocky coast and out into the ocean, happy for being lost in so much blue.

. . .

Pettit pushed different envelopes, after he found work at the Los Alamos National Laboratory in 1984. Today's Los Alamos, New Mexico, looks like any other small mountain town; they play high-school football and drive up to buy cheeseburgers from Sonic. But behind that anywhere façade, there lurks a darker history. During World War II, Los Alamos was a closed camp, guarded and gated, the most secure place on earth. A team of top military scientists, led by J. Robert Oppenheimer, were secreted away there, and together, they learned how to split the atom. Next they learned how to turn the desert into glass and make clouds shaped like mushrooms, and the bulk of what they learned was soon dropped on two cities in Japan, Hiroshima and Nagasaki. In the years of peace that followed, the scientists who remained behind became part of a larger government laboratory. In it, work continued apace on making and maintaining bombs, but Los Alamos also became a larger mecca for the microscope set. More than 14,000 employees now come in from as far afield as White Rock, Santa Fe, and even Albuquerque. It has become a place where chemists and physicists and engineers are free to explore every corner of the universe. It is a lab without limits.

When Pettit first came aboard, he was assigned to the Dynamic Testing Group, which meant that he toyed with detonation physics, making conventional high explosives. But true to form, and with the encouragement of his superiors, he started to develop his own research programs on the side. He jumped from project to project as fast as he could dream them up, confined only by the number of hours in a day. Even after he was told to go home, he would work away in his garage, which, after driving through a blizzard to an otherwise unattended auction, he had filled with surplus from the lab. It came to look as though it had been pulled out of a cheap science-fiction serial or a comic book. His garage even boasted a collection of three-phase tools, which required Pettit to sneak out and tap into his neighborhood's electrical grid to power them. He might have risked casting a good chunk of Los Alamos in darkness had he not been working since childhood on his touch.

In particular, his skill at building instruments soon earned him a spot in the Earth and Environmental Science Group. First, Pettit found a way to sample and analyze the fumarole gas spewed out by active volcanoes. He would travel to lava-born places like New Zealand and tap what bubbled up from the center of the earth. Next, he began firing up sounding rockets to probe noctilucent

clouds—eerie, electric-blue clouds that glow brightest at night, usually over polar regions, and always on the fringes of space. (They
were first observed in 1885, about two years after Krakatau exploded and coated the upper atmosphere with a thin layer of ash.
But the clouds have inexplicably persisted and spread in the century-
plus that's passed since, and Pettit never could resist an unsolved
mystery.) Finally, he studied materials processing in reduced gravity.
That vein of exploration gave Pettit the chance to make several
flights on board the KC-135, the infamous "Vomit Comet" that
dives and climbs in a series of parabolas to simulate weightlessness.
Its passengers know something of what astronauts know; the feeling is as close to space as most of us will ever get.

But it wasn't close enough for Pettit. Even before he began
working at the lab, he had put in an application at NASA. After
more than six months in Los Alamos, he was finally flown down to
Houston for an interview but was ultimately rejected—despite the
fact that his father had delivered one of the nurses who had helped
evaluate his fitness. He remained optimistic, however, and every
year he kept his application updated. Every year he dressed up his
project résumé with more lines like "solved problems in detonation
physics" and "conducted atmospheric spectroscopy measure-
ments." He eventually won another interview in 1986, but he was
again rejected. A third interview and rejection came in 1993.

By then, it started to feel as though Pettit had become an experiment all on his own, measuring how much disappointment a man
could stand.

. . .

While Don Pettit contemplated throwing himself into one of his volcanoes, Ken Bowersox raced toward the edge of his dream. After
only eighteen months at China Lake, Bowersox was accepted into
NASA's astronaut corps in June 1987. Despite its seeming inevitability, the moment remained a jubilant one. Every year, more than one
hundred pilots—already culled from the best of the best by their superiors in the navy and air force—submit applications for an astronaut pin. No more than a dozen are interviewed, and, in a bumper

year, perhaps three will receive a phone call from the Johnson Space Center, telling them to pack up their things and head to Texas. But Bowersox's arrival in Houston was ultimately bittersweet. NASA was still reeling from the *Challenger* disaster the year before, still deep in its long recess of reflection and self-doubt. By the following August, Bowersox had completed his training, passed every evaluation, and was ready for flight, but the shuttle fleet was not. It wasn't until September 29, 1988, almost three years after the accident, that a short-staffed *Discovery* took an abbreviated trip into space. It was longer still before the shuttles were running at full capacity, and the astronaut backlog that had built up—new recruits are planted at the back of the flight selection line—meant that it was almost five years before Bowersox was finally weightless.

Whether by design or by fate, his abnormally long ground tenure gave way to the sort of career that seemed destined to end on station. For his first flight, in the summer of 1992, Bowersox piloted a specially equipped *Columbia* on the longest shuttle mission yet, STS-50. It had been equipped with the first United States Microgravity Laboratory and, more important, the first Extended Duration Orbiter, a collection of improvements that included additional hydrogen and oxygen tanks for power production, more tanks to pump nitrogen into the cabin's atmosphere, and a better system for scrubbing carbon dioxide out of the crew's air. Combined, the changes allowed for shuttle flights lasting more than ten days. In this, their inaugural test, they pushed things a little further than that, spending thirteen days, nineteen hours, and thirty minutes away. (In fact, the trip was a day longer than expected because heavy rain back at Edwards Air Force Base delayed the scheduled landing.) In less than two weeks, Bowersox had gone from Houston greenhorn to the veteran of 5,716,615 miles in orbit.

A little more than a year later, in December 1993, he again piloted a high-profile mission, this time on *Endeavour*. STS-61 was devoted to repairing the Hubble Space Telescope, which, after much promise and anticipation, had delivered pictures that looked more like rain-streaked windows than anything that resembled the heavens. After a successful capture and a record five space walks lasting more than thirty-five hours (none by Bowersox), the Hubble was saved. Solar arrays, gyroscopes, an improved planetary camera, and a system of mirrors were all installed to fix the telescope's power, pointing systems, and focus. Aside from rescuing NASA's reputation, the mission also boosted confidence that the shuttle and its crews could help construct and maintain the embryonic International Space Station. It was the sort of accidental first step that made crawling seem obsolete.

In the fall of 1995, Bowersox made his own graduation, this one to the rank of commander, returning to helm *Columbia* on STS-73. The mission did not start well: its launch was scrubbed six times to tie the record for prelaunch jitters set by STS-61-C. Glitches included a main fuel valve leak, hydraulic problems, a failure in the main engine controller, and a minor meteorological inconvenience called Hurricane Opal. But *Columbia*'s belated tour was a success, as well as Bowersox's longest flight, at fifteen days, twenty-one hours, and fifty-two minutes. He also threw out the ceremonial first pitch for the World Series from space: a giant television audience watched him toss a ball that ducked out of camera range before falling out of the night sky into Cleveland's Jacobs Field.

Most recently, Bowersox had commanded STS-82, strapping into the front seat of his third shuttle, *Discovery*, in February 1997. For the second time, he helped capture and repair the Hubble telescope. Another five space walks prolonged its life span; Bowersox demonstrated the fine touch he had acquired at the controls by boosting the telescope's orbit, too. By the time he succumbed to gravity's call for the fourth time, he had logged a total of fifty days in space and rocketed across more than 23 million orbital miles. Given that rookie astronauts were forced to wait years for their first flight—just as he had been—Bowersox had become the sort of man who walked down the halls in Houston trailed by stares and whispers.

. . .

Pettit also heard his name whispered, but for different reasons.

In April 1996, he was punching his version of the clock: living on a small boat with five other scientists from the Los Alamos National Laboratory, anchored back in New Zealand's Bay of Plenty, collecting gas samples from the White Island volcano. The previous autumn, he had traveled to Houston to interview for a fourth time with NASA. In the months that had passed since, it had become harder for him to hold out hope that this time he'd clinched it. Occasionally, most often late at night, he would wonder what if, but now he was more occupied with his work and trying to beat down a sinus cold that was making his life miserable on board that tiny boat.

Things got worse when a storm began blowing in. Soon the waves were too large for the men to parry; they would need to raise anchor and head for shore. By the time they made it to the seaside town of Wakatani, almost all of the accommodations had been booked for the night. The six scientists, Pettit included, squeezed into a small cottage, rolled out their sleeping bags on the floor, and tried to shake their fevers and chills. They had just drifted off when the phone rang. It was two o'clock in the morning, and it was Houston on the other end of the line. Pettit tried not to sound as sick as he was.

"Are you still interested in becoming an astronaut?" a smiling voice asked him.

"Yep," he said.

And that was it. Pettit had made it, fourth time lucky. He tried to get back to sleep, but the elation of the moment—not to mention his awful cold and the sound of the rain beating against the cottage windows—made shut-eye impossible. Still, it took until morning's first light for him to realize that the conversation hadn't been a dream and that the course of the rest of his life had changed with a single phone call. He would be leaving the lab, selling his house, packing up Micki and their three dogs, and moving to Houston. He would become an astronaut. He would fly in space.

But first, he would feel like a freshman on an unfamiliar college campus. Houston was new. He didn't know his way around this sprawling, landmarkless city or where the nearest grocery store was or where the good restaurants were. Newer still was his job, far from your usual office transfer. Only a few days after arriving in town, he was shipped off to Pensacola, Florida—nearly two decades after Bowersox had passed through—to complete ground survival training, just in case the T-38 he would be mentored in went down. He sat in shuttle simulators and marveled at how different it felt from sitting in an airplane. He pulled on a big white spacesuit and jumped into the world's largest swimming pool, feeling more like a manatee than an astronaut.

Through it all, he did his best to fit in. He did his best to feel like part of the gang. In a lot of ways, however, he remained a man apart. He had begun unpacking his tools and lab surplus into his new tricked-out garage, and a curious Pettit—busy exploring his new universe—had taken it upon himself to stir up a bowl of liquid oxygen, one of the principal agents of rocket fuel. Now, sitting at the back of a crowded class during a lecture on propellants, Pettit shot up his hand when the talk turned to the very liquid oxygen that he had stored where most men kept their hedge trimmers.

"Do you know what color liquid oxygen is?" Pettit asked when called upon.

"Well, no," the lecturer said. "It's not really the kind of thing you can take a peek at."

"Well, um, I know," Pettit said. "I just made some. It's blue."

The rest of the class turned around in unison and stared hard at him, as though he'd just farted. It reminded him of the way his schoolmates had stared at him when he started that fire back in Tucson.

This time, Pettit stared back. "I just thought you'd be interested to know," he said.

In that moment, his reputation in the astronaut fraternity was sealed, probably forever. Even in a class filled with extraordinary men and women, even among dozens of astronauts as accomplished as Ken Bowersox, Don Pettit stood out. It was clear from the beginning that he would chart a different course. It was clear that he was something like a satellite, on an orbit all his own.

. . .

And yet, for all their differences in personality, for all their divergent history, Pettit and Bowersox would make for a seamless team. On the surface, they looked like the oddest couple, the pilot and the scientist, the arrow and the archer, the veteran and the rookie, the first-stringer and the reserve, blue eyes and brown, short and tall. They were a bad buddy cop movie come to life. But in the end, they had enough in common to find ways to tie themselves together; in them, somewhere, was the foundation for an impossible-seeming union.

Not surprisingly, neither can remember when or where they first met. Their memories of each other begin with time spent in the pool, training for weightlessness. Although their paths might have run next to each other long before that dip—somewhere out in the desert, or in the Florida panhandle—they probably first saw each other in the crowded halls of the sixth floor, Building 4 South, at the Johnson Space Center. Even then, even if they had bothered to make eye contact or perhaps nod and smile, they never would have dreamed that they would come to live together in space. It would have been harder still for them to imagine that one day they would cry together over the loss of shared friends and a space shuttle. How much would need to happen for that to happen, too?

Against very long odds, it did.

The simplest explanation is that, from childhood, Bowersox and Pettit were both drawn to space as though it were a light-filled window. That sort of yearning must spring from the same place, a place deeper than even John Glenn's voice. Perhaps their collision was born of their heading for the same destination.

Perhaps, except that every astronaut bunked down in Houston has a different motivation. For some it's the thrill of it. For others it's the glory. For many of them, it's pulling on an orange spacesuit that gives them an almost electric charge. For guys like Bowersox, it's the flying. For the rest of them, for guys like Pettit, it's the possibility of finding out something new about how the universe works that gives them goose bumps. For each of them, the most honest reasons for having landed here are as varied as what they like to eat or how they like to spend their Sunday afternoons. Their occupation is not what binds them, and it's not where they're going that matters.

What counts are the places they've been and the places they're from. The landscapes of their hometowns might seem as different as their faces, at least on the surface—the heartland and the harbors, the flats and the mountains, Indiana and Oregon. But their geography is the same: all of them are from the places that call to people who know what it means to be alone. They come from our empty places, our hidden small towns and the folds in the map, as far as you can go away and still be home.

Of course, there's another, better reason why astronauts are born lonely. City kids don't have the room nor any need to dream. The lights and chaos burn away their imaginations. The only decent dreaming gets done out here, in our wider landscapes, in our deserts and canola fields, those beautiful places where we don't even have to look up to see all of the sky at daybreak and every last star at night.

4 TIME AND DISTANCE

There are wide-open spaces in Russia, too. Nikolai Budarin had looked up at the same stars that Don Pettit and Ken Bowersox had taken in like breath. He had imagined the same journeys, dreamed the same dreams, and now here he was with them, in space, only three decades after their two countries had raced for the moon. In the middle-aged lifetimes of Expedition Six, a seemingly impassable distance had been closed twice over.

Once, the Iron Curtain might have seemed the greater divide. The exploits of the Soviet space program were as much rumor as fact, secretive enough for officials to fail to include the Kazakh town of Tyuratam, home to 50,000 inhabitants and the Baikonur Cosmodrome, in any official census. Though already in the middle of relative nowhere, the *Soyuz* launch site had been made even more remote by the loss of its anchor: the town was even rubbed off maps, with just another uninterrupted vista left in its place. In the absence of meaningful satellites, Tyuratam was invisible to the outside world. It was the ghost at the heart of the machine.

Under the cover of that darkness, pulled back only by the periodic flashes of rocket boosters, the Soviets had recovered from their loss in the lunar race and would even begin pulling ahead. The next frontier—the turning of space from destination to colony—would become their dominion. The Americans had conquered distance and, with that, seemed satisfied by their temporary stewardship; the Soviet view of the universe looked a lot more like rent-to-own. The Soviets aimed to conquer time.

Their crusade would prove the more difficult one, perhaps because it offered no easy finish. There was nothing finite about their goal, no endpoint looming on the horizon, no unexplored ground to stick a flag into. They could never say they'd done it, pack up, go home, and have a parade. But in the process, they learned something that the Americans wouldn't catch on to for decades: the most important journeys and dreams are those without end.

. . .

The beginning was Salyut, the first of seven manned stations that the Soviets rushed into orbit between 1971 and 1982. It was an almost unqualified success, at least until it came time to bring home its three-man crew. Georgi Dobrovolsky, Vladislav Volkov, and Viktor Patsayev had become that rare combination of hero and celebrity during three fantastic weeks in June 1971. Their playful weightless exploits—including a seemingly insatiable appetite for somersaults—made for nightly viewing across the country, reality programming with an all-time great payoff. Their success helped restore Russia's collective faith in itself after the Americans had danced across the moon. They were exactly what the Soviets needed to see in themselves. They were triumph.

But all of that good feeling was lost when a recovery team reached their blackened *Soyuz* capsule on the Kazakh steppes and found disaster. Somewhere along the way, Dobrovolsky, Volkov, and Patsayev had died. Early fears that their long-duration mission had weakened them past the point of safe return proved unfounded. Their deaths were the result of a more mundane misfortune. The heroes suffocated when a broken valve leaked every last drop of their atmosphere into space. There was evidence that they had lived long enough to try to reverse their fate, that they had felt the life hissing out of their spaceship and tried to plug the hole. But the harder evidence revealed that they had run out of time. One by one, their bodies were laid out in the tall grass, and plans for a celebratory national holiday were canceled for mourning.

Years of gloom followed. Salyut 2 was lost shortly after it was launched in July 1972. Never having been occupied, it fell out of the sky after it was punctured by debris when its delivery rocket exploded. Although a truthful history has been lost in the mire of Soviet misinformation campaigns, it's believed that two other failures followed. Finally, what came next was worst of all: the Americans prepared to send their own station into space, and in Moscow, there were fears that they would use it for more than somersaults.

. . .

In reality, the star-crossed Skylab was a halfhearted effort at a semi-permanent space colony, pulled together using hardware left over from the *Apollo* missions that never were: *18, 19,* and *20,* each submerged under the rising national sentiment of been there, done that. The third stage of a mothballed *Saturn V* rocket was slung into orbit with the hope that a series of crews would occupy it, learning a little of what the Soviets already knew about bunking down in space rather than just passing through it.

The rocket's shell was fitted out and launched unmanned on May 14, 1973; not much went smoothly after that. When Skylab reached orbit 270 miles above the earth's surface, Mission Control discovered that the module's meteor shield had broken off, which was bad news on a couple of fronts. First, the shield was designed to help shade Skylab's workshop from the heat of the sun. Without it, the crew inside would slow-roast at 250 degrees Fahrenheit, a temperature twice as hot as those felt on Death Valley's salt flats. And second, the shield took one of Skylab's two power-generating solar panels into oblivion with it. The remaining debris had jammed Skylab's second and only remaining panel, preventing it from deploying properly, like a bird's broken wing.

Undeterred, NASA chased Skylab's first crew after it, only eleven days later. Instead of conducting the full range of planned experiments, Pete Conrad (a veteran of *Apollo 12*), Paul Weitz, and Joseph Kerwin were assigned the difficult task of making the ailing station habitable. Under challenging circumstances, they did some kind of job, unfolding and attaching a makeshift tarp to replace the lost meteor shield as well as clearing the station's lame-duck solar panel of debris and popping it into place. That gave them just enough power to head back inside and hang out for the better part of a month—for twenty-eight days, then the record for space endurance.

By the galaxy's two-star standards, Skylab's crew enjoyed palatial surroundings. Unlike more modern stations, which have been modeled after factories, Skylab was a place for living, always more of a home than an office. There was a collection of sleep compartments tucked away in a quiet corner; a ward room with rows of food-storage lockers and chillers and a window for looking out on the earth; even a collapsible shower, the sort of decadence that Bowersox, Pettit, and Budarin did without on the International Space Station. A big vacant attic also offered Skylab's crews a wide-open respite, perfect for whenever they needed some elbow room or felt like honing their zero-gravity acrobatics, just for kicks.

But such luxuries also betrayed a certain cultural weakness in the Americans. The fact was, and remains, that men born and raised in austere, cramped Muscovite apartments adapt more easily to living in austere, cramped space stations. The Soviets were harder. For instance, unlike astronauts, cosmonauts have always refused to wear diapers during flight. They would rather starve themselves in the days before liftoff and flush their pipes with an ice-water enema than get caught wearing a nappy. That hairy-assed triumph of the will, combined with a societal emphasis on the needs of the community over the desires of the individual—built on that old hammer-and-sickle platform of self-control and sacrifice—has served them well in space. Don't forget, too, that a country in which smiling is viewed as a weakness has never had trouble finding men cold-souled enough to go months without hugs. First the Soviets, and now the Russians, have almost been bred to live year-round in personal winters.

The glad-handing Yankees, it seemed, not so much. After Skylab's first crew made their safe return and its second—Alan Bean (Conrad's crewmate on *Apollo 12*), Jack Lousma, and scientist Owen Garriott—spent an uneventful fifty-nine days in orbit, Skylab's third and final crew put their finger on the seeming limits of American endurance.

On November 16, 1973, Gerald Carr, William Pogue, and Ed

Gibson lifted into space. They were all rookies, and the novelty of the place carried them through their first weeks. They took photographs of Comet Kohoutek, and Gibson profiled a solar flare, and they began what would eventually total more than twenty-two hours of near-perfect spacewalking.

But after the honeymoon, the breakdown began. They started to spar with the ground over a litany of complaints, from the poor quality of the towels to the awkward placement of the toilet, which too often turned morning dumps into hand-to-hand combat. The ground got cranky in turn, issuing a sharp reprimand when Pogue lost his lunch and Carr, instead of going by the book and bagging the puke for future analysis, flushed it. "We won't mention the barf," Carr said to Pogue, unaware that Mission Control had been listening in on the entire episode.

That rebellious streak exploded into full-blown mutiny toward the end of the crew's sixth week in space. Complaining of overwork and a lack of cooperation from Houston—as well as those lousy goddamn towels—Carr, Pogue, and Gibson staged a distinctly un-American one-day strike. Though they eventually went back to work and spent eighty-four days in space, setting NASA's latest endurance record, the crew's legacy and the future of long-duration American spaceflight were left clouded. Before their return on February 8, 1974, the crew was told to boost Skylab into a higher orbit than usual. There, it would be left to hang as if from a string, powered down and dormant.

It stayed quiet for four years, until 1978, when it was revived from the ground just in time to chart its months-long fall to earth. NASA tried and failed to control its descent, and the public began taking a dim view of the nutbar scheme, nonplussed by the idea of space junk falling on their heads. Fortunately, when Skylab finally did make its fiery plunge on July 12, 1979, the only surviving fragments tore into empty patches of Australian desert.

Despite the program's successes—a better understanding of the physical and psychological effects of calling space home, for starters—Skylab's Chicken Little return saw it finish on a cracked note. NASA was left cornered by those memories, forced to turn its attention to the shuttle and quick dashes into space. The American endurance record set by Carr, Pogue, and Gibson would last for more than twenty years. Even then, it would take help from the Russians to break it.

. . .

And yet, in a strange way, the Americans had first helped the Russians: by beating the scientists at Star City so plainly, in both conquering the moon and now turning their attention toward space, Houston's technocrats had inadvertently freed their competition from the burdens of coming in first. It was as if some invisible pressure valve had been released, giving the Soviets the chance to catch their breath and redouble their efforts. They went back to work in a calm, calculated, and ultimately enviable way, without the artificial weight of deadline or the specter of shame. For the first time since *Sputnik*, they weren't forced to race against the clock; instead, they concentrated their efforts on calendars.

The result was much-needed success, foreshadowing the greater glories that were to come. Salyut 3 and Salyut 4 were both slipped easily into orbit and visited by five different crews through 1974 and 1975. (Each did, however, experience at least one link-up failure.) With every launch came important advances. Cosmonauts became better prepared for the physical and psychological demands of long-duration flight. Their vessels were also made more ready; Soviet engineers brainstormed their way toward a kind of inventor's immortality, their gifts still giving today. Among other things, they devised an air lock to jettison trash, a water purifier that recycled moisture collected out of the air, a zero-gravity exercise bike, and a small vegetable garden called Oasis. The developments allowed Salyut 4's final visitors, Pyotr Klimuk and Vitali Sevastyanov, to spend sixty-three days in space—battling a green-mold epidemic and humidity high enough to fog their windows, but also nearing the endurance record set by Skylab's rowdy last crew. They had come close enough, in fact, to pull the Soviets back to even, close enough for the gap to be bridged in the most palpable way.

Like all thaws, the melt that finally pooled the American and Soviet space programs was a slow one. Its first trickles, surprisingly enough, came in the days before the space race kicked off in earnest. During his inaugural address in 1961, President John F. Kennedy advocated a shared journey into space: "Let both sides seek to invoke the wonders of science instead of its terrors. Together let us explore the stars." Come the Bay of Pigs, the Berlin Wall, the Tet Offensive, and 100-megaton nuclear tests, terror became the lead horse. Mercury and Vostok became stunt doubles for Kennedy and Nikita Khrushchev; Gemini and Voskhod stood in for Lyndon B. Johnson and Leonid Brezhnev.

Victory ended the battle. It helped that the winner was gracious. Along with the flag and a patch to commemorate the lives of the *Apollo 1* astronauts, Neil Armstrong and Buzz Aldrin left behind a pair of medallions in tribute to Vladimir Komarov and Yuri Gagarin in the moon dust. Komarov died in April 1967, when the first Soviet *Apollo*, *Soyuz 1*, hurtled to earth, its parachute lines tangled. Gagarin, the first man in space, had died in a mysterious jet fighter crash in March 1968. In such tragedy, there was unity. The patch and medallions were like the finishing touches on a sad song for which both sides had written verses.

Three years later, they were brought that much closer together by a thin document held in a plain blue binder. Signed by Brezhnev and Richard Nixon during the first American-Soviet summit in May 1972, the agreement spelled out the shared desire to witness astronauts and cosmonauts shaking hands in space within thirty-six months.

Only slightly behind schedule, the Apollo-Soyuz Test Project saw a once impossible dream come true. On July 15, 1975, *Soyuz 19*, carrying Alexei Leonov and Valeri Kubasov, blasted off from the formerly invisible Baikonur Cosmodrome. A little more than seven hours later, Tom Stafford, Deke Slayton, and Vance Brand lifted off in their *Apollo* capsule. The two ships would soon become one, albeit in an ungainly embrace, looking a little like two insects joined at the head.

As unpretty as it would look, that union represented more than a feel-good photo opportunity. It was more than a moment. To see it happen, each side was forced to share technical information about their docking mechanisms, their communications and guidance systems, their flight control procedures . . . A long list of top secrets would open wide along with that hatch.

On July 17 came the historic announcement from ground control: "*Apollo*, Houston. I've got two messages for you. Moscow is a go for docking. Houston is a go for docking. It's up to you guys. Have fun."

The two vessels joined 140 miles above the earth. *Soyuz*'s oxygen-nitrogen atmosphere mixed with *Apollo*'s pure oxygen, and Stafford met the Soviets with a handshake. His was the first word between them: *tovarich*, the Russian word for *friend*. After more pleasantries, including an exchange of flags and plaques, the five men circled the earth for forty-seven hours. They ate together, worked together, dreamed together. And then it was time to say goodbye.

Soyuz 19 returned two days later. The last *Apollo* remained in orbit for three more days, splashing down in the Pacific Ocean on July 24, 1975. With the space shuttle's development under way, the plan called for Americans never to fly in a capsule again. It was the end of an era.

That fall, the world seemed at the end of another. The astronauts and cosmonauts toured each other's countries, warm and friendly, smiling for cameras and waving at crowds. Wonder had made its comeback down the stretch.

. . .

In June 1976, almost a year after rivals had become temporary friends, the Soviets launched Salyut 5. It marked the start of their steepest learning curve yet. Through successes and failures, they began to get a feel for the rhythms of long-duration missions, both in their men and in their machines. After Salyut 5 had run its course—longer than expected, because cosmonauts were getting better and better at maintaining a ship in orbit—Salyut 6 went up, and in that barrel-shaped home, the Soviets saw the benefits of their new wisdom fully realized.

The first assigned crew failed to link up, but the second crew, Yuri Romanenko and Georgi Grechko, made it aboard, the start of a planned ninety-six days in space. That record duration required the men to assume the role of living, breathing experiments. It also necessitated a number of technological developments, such as a *Soyuz* exchange program, because their capsule wouldn't survive as long in space as the crew might. Also, another docking port had been added to host the new *Progress*, the unmanned supply freighter still in use today.

The *Progress*, in both its design and length of service, is a prototypically Russian vessel. Essentially, it is a hollowed-out *Soyuz* capsule, fired into space on top of the same booster rocket used by its manned cousin. Where the crew normally sits, there is instead room for more than 3,700 pounds of supplies. In a separate compartment, spare tanks can be filled with propellant and water and pumped into another vessel or, today, into the International Space Station's own stores. A third module contains the ship's electronic equipment and sensors, the operation of which is entirely automated, including a radar system called Kurs that guides the *Progress* into its designated docking port. Once in place, the ship's hatch can be cracked open by a joyous crew, hopeful that at least a few of those 3,700 pounds consist of fresh fruit, chocolate, and spices. (Once it has been emptied, the *Progress* is filled with trash and returned to expire in the earth's atmosphere, burning up somewhere over the Pacific Ocean.)

Romanenko and Grechko were especially thrilled to see their own chunky *Progress* appear outside their window. Trying to get some flavor to penetrate their packed sinuses, they had burned through their entire condiment supply in the first five weeks of their mission. Happily, they accepted a special delivery of mustard and horseradish, bushels of scurvy-preventing apples and oranges, and beef tongue in jelly, a Russian delicacy, washed down with apricot juice.

A "psychological support group" on the ground oversaw the care packages, adding mail from home, newspapers, and even a guitar to the mix, giving the men something to listen to other than the chattering of instruments. The shrinks also encouraged the crew to keep tending their onboard garden, not because anything edible was coming out of it but because the pair seemed bolstered by the presence of green and sprouting things. (To make them feel even more at home, hardwood paneling had been installed inside their quarters.)

Keeping Romanenko and Grechko up physically was trickier. Strenuous workouts were made part of their routine, including running on an improved treadmill, which, like *Progress*, remains part of the program today. Their relationship was also strictly supervised. They had been told to monitor each other's moods, to try to help each other with difficult tasks, to learn when the other man was sending out signals that he needed a shoulder to lean on or, more often, to be left alone. Given their living and working sometimes literally on top of each other, they did remarkably well. Their only suggestion to the ground was that separate sleeping compartments might be a nice touch in future stations. The ground agreed, and in the meantime, colorful partitions were sent up along with more spices and fruit. Each new lesson was layered on top of the last until, come the end of their mission, Romanenko and Grechko had virtually mastered the hard art of isolation.

Subsequent crews lasted longer on Salyut 6—Vladimir Kovalenok and Sasha Ivanchenkov stayed for 140 days in 1978, and Valery Ryumin and Vladimir Lyakhov spent 175 days up there the following year, breaking any previous endurance record and finally doubling the best effort of the Americans—but in some ways, new ground would never again be broken. By the time Ryumin went up for a second mission, spending another 185 days in orbit, just about every important hypothesis had been proved. Men could live and work in space, in harmony, for a seemingly infinite amount of time, without turning to dust or going insane; they could refurbish their homes on the fly and extend their life spans, if not indefinitely, then well enough to make their space stations track more like stars than meteors; and they established, once and for all, that men were not limited to exploring space like skin divers holding their breath. They could *inhabit* it. In that way, they forever raised our ceilings. They made even Mars seem possible.

. . .

All that good was undone when the USSR invaded Afghanistan in 1979, leaving the 1980s almost as cold on earth as it was in space. The Soviets pushed on with their Salyut program, sending six long-duration crews to the seventh and largest version of the ever-evolving station, moving steadily toward a continuous manned presence in space. (The three-man crew of Leonid Kizim, Vladimir Solovyov, and Oleg Atkov set the latest endurance record, spending two-thirds of 1984—an astonishing 236 days—in orbit.) The Americans, meanwhile, launched their gleaming new space shuttle *Columbia* in April 1981, ignoring for the moment that it had nowhere to shuttle to. The twin efforts, in many ways, mirrored the steep ideological divide between the two countries. The Americans darted into space like fresh-faced tourists and looked good doing it. The Soviets kept pumping old, ugly technology and their army of unsmiling cosmonauts into orbit, firm in the belief that one of these days, they would figure out a way to make them stick.

They did. While the shuttle was essentially relegated to the role of satellite delivery service, the Soviets unveiled plans for Mir, the world's first permanent multiple-module space station. It was more than just an outpost; it was a home.

The first module was launched successfully on February 20, 1986. Seeing as the Russians have never gone in for change, its design was identical to the trusted Salyut capsule, with the notable addition of several more docking ports for subsequent modules or pit-stopping spaceships. The base block, as the original module was known, was soon joined by Kvant, a full-fledged astrophysics laboratory, stuffed with telescopes and instruments. Its docking was delayed by a bag of trash that had somehow found its way outside and into the designated port. In their no-nonsense fashion, two cosmonauts suited up, went outside, fumbled around in the dark, and pulled the bag free.

Kvant 2 went up in November 1989. Most important, it contained a new toilet, allowing the cosmonauts to all but abandon the original crapper, which sat two feet from their dinner table. The fourth and final planned module, Kristall, was tied to the end of the train six months later. With its solar panels and unflashy architecture, the completed station looked like a dragonfly, large enough to swallow six men whole.

Its size and scope allowed for grand new visions. In 1991, succumbing to the spirit of glasnost, Mikhail Gorbachev and George H. W. Bush agreed to a ceremonial swap of spacemen. A NASA astronaut would visit Mir; a Russian cosmonaut would take a spin on the space shuttle. August's coup attempt and the subsequent collapse of the Soviet Union in December put those plans on hold. The chaos also left a boy-faced flight engineer named Sergei Krikalev stranded on Mir.

· · ·

In May, he had waved to his family one last time through the window of a white-and-yellow bus and been driven away, along with his commander, Anatoli Artsebarski, and a British researcher named Helen Sharman. That night, in the sleepless hours before launch, they had climbed to the roof of their quarantine hotel and raised Mir with a handheld radio, pointing an antenna at the sky. Voices had crackled through a wash of static.

The following morning, already suited up, they had boarded another bus, this one white with blue trim. Their rocket had already been pulled to the launchpad on top of a giant train car, where it waited for them, gray and shining, its white nose cap glowing in the sun. On their short drive over to it, they had watched final farewell videos from their families on small monitors. They had also been given bundles of fragrant wormwood twigs, a goodbye tradition, which they had pressed to their noses. Krikalev had lingered over them for an especially long time, breathing deep.

Their liftoff had been perfect. A little more than two days later, they had docked with Mir. Watching film of it, with the right soundtrack, an audience might have confused the maneuver with ballet.

The two cosmonauts inside, Viktor Afanasiev and Musa Manarov, had greeted their visitors with bread and salt. Krikalev had seemed overwhelmed during the changeover period, those hectic days when all five cosmonauts had remained on Mir. He was often lost in thought and quiet, staring out the window. Earth looked bright and breathtaking from such great heights. All of the planet's small erosions and scars were invisible. There were only continents and oceans, wrapped in clouds.

Afanasiev, Manarov, and Sharman had returned home, leaving Krikalev and Artsebarski alone in their new world. They had worked in increasing clutter, repairing aerials and docking systems. They had also conducted a long list of experiments. In the downtime, they had listened out for word from home and occasionally spoken with their families. Krikalev had learned that his wife had bought new furniture for their apartment. His boy had gone swimming one day, fishing the next.

Then came the coup attempt. They heard about it on their amateur radio, the same one they'd raised from their hotel, but the short bursts of news did little to convey the reality on the ground. They couldn't see the violence that came on the heels of darkness, the tanks smashing through roadblocks and over trolley cars, and they couldn't hear the sirens and the screams, the bottles breaking on the rain-slicked pavement. Even through night's prism, they couldn't make out the fires. Suddenly they felt very far away.

The scale of change didn't truly hit them until the fall, when a replacement crew was scheduled to arrive at Mir. Artsebarski would be switched out for Aleksandr Volkov, commander for commander. But instead of including a flight engineer to take Krikalev's place, Star City sent an undertrained Kazakh, Toktar Aubakirov, and an Austrian researcher named Franz Viehbock along for the ride. The Austrian's place had been guaranteed by a big check. Aubakirov's inclusion was political, a concession to a new reality. With the Soviet Union on the verge of collapsing in on itself, there were fears that the increasingly sovereign nation of Kazakhstan might deny Russia access to the Baikonur Cosmodrome and thus to space. So, just to be safe, Aubakirov would jump the line and enjoy a short stay on Mir—keeping the Kazakhs happily receptive—and Krikalev would see his tour of duty extended indefinitely. After a tense changeover period, Artsebarski high-fived his friend and ducked out of Mir, taking the Kazakh and the Austrian back to earth with him.

In the quiet that followed, Krikalev pressed his new commander for news.

The ground also passed tidbits along: "There's a mixed international company talking to you, from a brand-new control center with a brand-new map. The Baltic States have already got a different color. And the Kuril Islands are preparing for a change of color, too."

Gorbachev made way for Boris Yeltsin, the hammer and sickle for red, white, and blue, Leningrad for St. Petersburg.

Finally, in March 1992, Krikalev's belated ride home rocketed toward Mir. He packed his things, listened to the radio, and stared out the window some more.

"Last year, you left the Soviet Union," a reporter said to him from the ground. "Now, you return to Russia. How do you feel about such drastic changes?"

In response, there was only silence.

On March 25, a record 310 days after waving goodbye to his family through the window on a bus, the last Soviet citizen was lifted out of his scorched Soyuz capsule. Although he had run on a treadmill for two hours every day, Krikalev was too weak to walk or stand. He was placed in a chair and carried like a sultan to a recovery tent. His original commander, Anatoli Artsebarski, was waiting there with sunglasses. He feared that the bright colors Krikalev had seen only through the filter of space might now hurt his eyes.

Still in his blue flight suit, he was bundled into a plane and flown back to Moscow. He smiled when the flight attendant came by with her drinks cart.

Krikalev's first taste of his brave new world was a can of Coke.

. . .

Maps and soda aside, the blurring of lines continued. In June 1992, to liven up their lackluster summit, President Bush and Boris Yeltsin came to yet another space cooperation agreement. It was decided again that an American would visit Mir, but this time around, two Russians would fly aboard the shuttle. After Bush lost that November's election, political upheaval once again forced a change in plans.

Not long after Bill Clinton had won the White House and begun reviewing the books, his administration coldcocked NASA with a 20 percent budget cut. That kind of money couldn't be found by cutting back on paper clips and magazine subscriptions. Something real had to disappear. And there was only one line in the budget that could be hacked out without killing the rest of the leviathan: Space Station Freedom.

Nine years after it had been proposed by Ronald Reagan— "America has always been greatest when we dare to be great," he had said in his launching of the project—NASA's late answer to Mir had morphed from white star to white elephant. There wasn't a single piece of hardware to show for the $8 billion that already had been dropped on it. And with $23 billion in projected additional costs, the price of Freedom was, for the first time in the history of the American space program, just too high. Overnight the blueprints were torn up and tossed away, and 20,000 engineers and technicians suddenly lost their footing. By late afternoon the next day, a sense of united purpose had given way to shock.

In the frantic weeks that followed, scaled-down redesigns were floated in the hopes of salvaging something from the wreckage. They weren't successful until Russia's star chasers, as though on cue, found the entrepreneurial spirit that only its Mafia had yet tapped. Maybe, they suggested coyly, a new space station was something the two countries could build together: the cash-strapped Russians, who had taken to working with the lights off to save money, would bring their long-honed but underfinanced expertise; the Americans would bring most of the bankroll.

Of course, the trade was a little more complicated than the usual fee-for-service. Building a space station is a communal exercise, which the Russians knew something about. It's also an expensive one, which the Americans had learned only too well. That was all true enough. But their money would give the Americans more than a light in the sky. It would also give them the leverage they needed to push the newly liberated Russians down a more righteous path than they'd seemed inclined to take on their own. Rather than hawking their nuclear secrets to India or Iran, Russia's fledgling democracy would be built on a more benign foundation. And the Russian engineers who had been sprung into underworld mercenary careers by their bouncing paychecks might now find reason to stay onside. What the Americans were buying themselves, in essence, was a tied-up bundle of guarantees, in space and on the ground. It was a honeypot deal.

And so, in September 1993, U.S. Vice President Al Gore and Russian Prime Minister Viktor Chernomyrdin announced joint plans for the International Space Station. Canada, Japan, the eleven members of the European Space Agency, and Brazil decided to play along, but the meat of the matter remained the same: the Americans would pay the Russians to help them build the station that had been too long in coming.

To fill the wait, the Americans handed the Russians a $400 million down payment, in exchange for training a corps of long-duration astronauts and, over the next several years, stowing seven of them on board Mir. Abiding by the original agreement between presidents Bush and Yeltsin, NASA would also play host to a pair of Russians on the shuttle.

On February 3, 1994, the first of those Russians tagged along for a routine mission on *Discovery*. He was none other than Sergei Krikalev, the flight engineer who was just getting his land legs back after his extended mission on Mir. Now he escaped to a place more familiar than this world had become. "There were those who said I should have stayed on Mir," Krikalev said. "Things were better up there."

A year to the day later, Vladimir Titov buckled into Krikalev's seat on *Discovery*, this time en route to Mir for a mock docking, one last rehearsal in the run-up toward ultimate union. With Jim Wetherbee at the controls—the same veteran astronaut who later shuttled Bowersox, Pettit, and Budarin to the International Space

Station—the Americans again demonstrated their mastery of distance. *Discovery* nearly kissed Mir, traveling thousands of miles before stopping within forty feet of its one-day destination.

Wetherbee had penned a speech to mark the momentous flyby. "As we are bringing our spaceships together, we are bringing our nations together," he read. "The next time we approach, we will shake your hand, and together we will lead the world into the next millennium."

Looking out at *Discovery* through his breath-fogged window, Russian commander Aleksandr Viktorenko was less prepared but no less prophetic. "This is almost like a fairy tale," he said. "It's too good to be true."

. . .

In February 1994, not long after Sergei Krikalev had made his shuttle flight, the first American astronauts began training at Star City, where a vast but now decaying complex sprawled out. There were gray apartment blocks and cottages set aside for the cosmonauts, ancient mock-ups sunk to the bottom of giant water tanks, *Soyuz* simulators, chalkboard-lined study halls, and gyms filled with treadmills and old-school exercise bikes. When this bleak campus wasn't bleached white with winter, it was a frayed patchwork of Russian institutional chic: everything was painted with the same dull slate of grays, greens, and blues. The washed-out palette added to the ill feeling of the place, that it was the home of so much faded glory.

Now it was home to a small band of Americans, too, about to embark on crash courses in all things Russian: language, technology, philosophy, food and drink. That vast program of study was made harder by deep freezes, culture shock, homesickness, and all of the small, wearing skirmishes that follow détentes. The power plays marked the start of a long feeling-out process between old enemies who were not yet friends.

The Russians were difficult to get to know, resistant to Houston's wide smiles and firm handshakes, but with time, small truths trickled to the surface. While the American astronauts were individuals, relative freethinkers and independent spirits, the Russians seemed cloned from a single prototype. They yielded to authority and obeyed without question, willing to take unreasonable chances so long as they were ordered to. With less emphasis on the individual, there was less value placed on an individual's life, and cosmonauts had been turned into fatalists in the most literal sense. Many of them believed in their hearts that they would one day die in space, and they had become almost mechanical in their march toward that grim destiny. One of the Western imports, Mike Foale, thought of his Russian colleagues as slaves. He used the word not lightly but a lot.

From that foundation sprang everything else. The Russians' appetite for doom sometimes fed the fearlessness they were legendary for, like those two spacewalking cosmonauts who had staked their lives on a thin tether to pick that bag of garbage out of Mir's docking port. But that machismo was also born of necessity. With long-duration flights, there wasn't a chance to bring things down to earth and study them and decide on an optimal course of action through diagrams and decision trees. That was the American way. The Russians just got it done because they had to. Otherwise they fell out of the sky.

That, in turn, fostered in them a certain roughshod practicality. If Star City's Mission Control became infested with mice, which it did, cats were brought in. When it was discovered that pens didn't work in zero gravity, with nothing to draw down the ink, the Americans spent millions developing a pen that could write upside down. The Russians packed pencils.

They were more willing to make do and to make do without, relying instead on an almost comical catalog of superstitions to carry themselves through. Their preflight routine remains cast largely by missions past, each new rhythm layered on top of the last. Cosmonauts visit Yuri Gagarin's office, frozen like a time capsule, and pay silent homage; each bus ferrying them to the launchpad has a horseshoe stashed in it, as well as those bundles of fragrant wormwood twigs; that same bus will stop to give the cosmonauts one last chance to climb out and water down the right rear tire, the way Gagarin had relieved himself before his historic first flight; in the night before their ceremonial pissing, they will have watched a film called *White Sun of the Desert* for the hundredth time, reciting every line from memory. No one can remember why they must watch this one movie, exactly, but they do, always and without argument.

...

In March 1995, Norman Thagard became the first American astronaut to live aboard Mir. Two new American science modules, Spektr and Priroda—late additions to Mir's design and products of the newly hatched plan for intergalactic cooperation—had not yet been added to the dragonfly's back, and Thagard found himself in an environment nearly as hostile as the vacuum around it.

Mir's walls were covered with equipment, strapped down or bundled into place, years of debris that had built up like newspapers in an old man's apartment. It could take hours to find something in the mess, a bizarre collection of the vital (laptop computers and tools) and the frivolous (movies and cassettes of Russian folk music, finger paintings sent up by schoolchildren, and a strange poster of a woman with her big dog). The cramped modules were connected by hatches just three feet wide, and even they had been clogged with dozens of cables, extension cords, and ventilation tubes. The sleeping compartments were small, worn-seeming, and barren, adorned only with shaving mirrors. They added to the feeling that something about the whole place was uncomfortably ramshackle. Two treadmills sat in the base block, and if a cosmonaut ran at a particular pace, the frequency of his rhythm could cause the entire station to oscillate, as happens when someone jumps into a canoe.

And Mir stank. Sometimes it smelled like a junkyard, thanks to frequent antifreeze leaks that left bubbles of coolant splashing into walls. More often it smelled like sweat. With no shower and a limited supply of water—so short that there were plans for the crew's own urine to be captured and distilled—Thagard and company could only wipe themselves down every three days, and T-shirts had to last them two weeks. To the horror of the Americans, the Russians had taken to lighting up cigarettes, more out of the desperate need for a change in whiff than for the nicotine buzz.

Not surprisingly, Thagard didn't thrive. As well as suffering through the anxiety of a series of technical glitches (a faulty freezer ruined most of his experiments, leaving him underworked), he experienced rapid and alarming weight loss. He shed seventeen pounds in his first six weeks in orbit, dropping him from an already skinny 158 pounds to 141. In an effort to counteract this decay, he was put on a rigorous exercise program. But when he started working out with rubber expanders looped around his feet, one of the straps came loose, and he whomped himself in the eye, damaging his cornea. By the time the shuttle *Atlantis* docked at Mir at the end of June, Thagard was ready to get back home. His 115 days in space had broken the American endurance record set by Skylab's third crew, but it had nearly broken him, too. He retired from NASA soon after his return.

The second of the Americans on Mir, Shannon Lucid, had a better go of things. She was the first to enjoy roaming around the two new U.S. modules—floating from the salt-stained base block into the fresh digs was like stepping out from shadows and into the sun—and her easygoing personality saw her shrugging off some of the problems that had affected Thagard to his core. Even bum plays somehow worked out: a six-week delay in her retrieval flight allowed her to set the world endurance record for a woman in space. (Between March and September 1996, she had spent 179 days gravity-free.) She also landed her smiling self on the cover of *Newsweek*. After a too-long absence, American astronauts seemed poised to return to the ranks of heroes.

Unfortunately, Lucid's replacement had trouble shouldering the role. The Russians had resisted the inclusion of John Blaha from the beginning. Their psychological testing, much more refined than the Americans' rudimentary understanding of what space can do to your head, revealed that Blaha was too dependent on outside support and too much of a perfectionist to find happiness inside fly-by-night Mir. The Americans had no one to replace him with, however;

the line of volunteers willing to spend eighteen months training in Star City was short. Blaha launched.

Almost immediately, problems arose. He clashed with his crewmates, commander Valery Korzun and flight engineer Aleksandr Kaleri. The simmering conflict was made worse by Blaha's poor Russian. He felt increasingly isolated. He also fell behind the optimistic science schedule that had been drawn up for him in Houston. Still learning their way, the Americans governed the lives of their astronauts by a high holy document known as Form 24, which structured the course of their mission, minute by minute, day by day. That's how shuttle flights had always been conducted. But over the course of their long-duration assignments, each of the astronauts began to grate under the thumb of the ground, Blaha in particular.

Trying to catch up, he began cutting back on his sleep, much to the concern of his crewmates, who didn't like the look in his tired eyes. Soon Blaha was getting only three hours of rest each night. With time, even those short spells became scattershot. Fatigue and stress combined to push Blaha into a deep depression.

Eventually he came out of it, after finding solace in his softening crewmates, English conversation on Mir's ham radio, and comfort in watching tapes of old football games that were sent up to him. That was enough to get him through to the other side, and by January, he had learned how to get along in space. It had just taken him time.

When Blaha reached the end of his mission, he returned safely to earth. Like Norman Thagard before him, he gave up on the astronaut business shortly thereafter. Although he had some good memories of his time in orbit, there were many more that seemed like bad dreams, and if he was going to have any chance of leaving them behind, he decided, he needed to make a clean break.

. . .

Blaha's successor, Jerry Linenger, would bring his nightmares back with him, mostly because they were easier to quantify. The first came on February 23, 1997, during one of Mir's cramped changeover periods. There were six men on board: Linenger, Korzun, and

Kaleri, as well as the newly arrived replacements for the two Russians (commander Vasily Tsibliyev and flight engineer Aleksandr "Sasha" Lazutkin) and a way-paying German, Reinhold Ewald. Even before then, Linenger had battled growing frustration. Like Blaha, he had had trouble building friendships with his crewmates. He had also begun looking sideways at Mir, which felt to him as though it was on the brink of collapse. Some 1,600 breakdowns would plague it over its lifetime, but Mir was beset by almost routine failure during Linenger's tenure. The coolant leaks worsened; oxygen generators conked out; condensation formed on the walls and behind panels; the power flickered off and so did the climate control system, spiking temperatures to 90 degrees; the main computers crashed; and communication with the ground turned sporadic and ratty.

But even the worst of those complaints was made to seem like a trivial inconvenience when an oxygen generator that was being changed in Kvant caught fire, spitting out a blowtorchlike flame and filling the module with white smoke.

"*Pozhar!*" someone cried out. Although Linenger's Russian remained limited, this word he knew.

Two members of the crew tried to smother the flame with a wet towel, but it, too, caught fire. Fed by the oxygen streaming out of the generator, the flame grew and turned blue. Liquid metal began floating through the module, threatening to spread the blaze like sparks blowing off a forest fire.

Korzun scrambled for an extinguisher. He pointed it at the fire and—nothing. It didn't work. Banging it around, he lost it in the smoke, which was turning from white to black.

That was it. The crew scattered, looking for oxygen masks and a way out. One of the *Soyuz* capsules was readied for launch. In the chaos, it was forgotten that the second was on the other side of the fire. Now a grim realization surfaced through the confusion: for three of the men, there was no escape.

Korzun, sweating and swearing, dug out another extinguisher from the piles. This one worked, and he emptied it in the direction of the fire. He didn't find all of it.

Now the generator was in the middle of the module, spinning, propelled by the flame like a rocket. The fire licked the walls, scorching them black and threatening to melt Mir's hull. If it burned through, the fire would have finally been snuffed out by the loss of atmosphere, the lives of six men gone along with it.

Beyond desperate, Korzun found a third extinguisher. This time, he caught just enough of the flame to put it out.

In the aftermath, the Russians told the Americans that the fire had lasted ninety seconds. Linenger said it was more like fifteen minutes. But for once, time didn't really matter. It couldn't be measured by a clock. It had been measured in heartbeats, and how few the crew had thought that they had left.

. . .

Linenger's second near-miss came less than two weeks later. Life inside Mir was just returning to normal. Korzun, Kaleri, and Ewald had dropped back to earth. Tsibliyev and Lazutkin were settling in nicely, at least until the Russians decided to run a test.

Mir, like Salyut, was supplied between manned missions by *Progress*. Its dockings had always been automated, but the necessary electronics—specifically, the Kurs radar system—were proving too expensive in the new Russia, especially given that they were used once and then burned into oblivion. The idea was hatched to make the process manual. The ground would guide the ship within striking distance of Mir; the commander inside would take over, calling out signals with his computer and controlling the rocket's thrusts with a joystick, ushering the ship the rest of the way. Tsibliyev would be the first to give it a shot.

Unfortunately, in the middle of the test run, with the *Progress* somewhere out there rocketing toward Mir, Tsibliyev's monitor—his eyes, in effect—filled with static. He was flying blind.

Lazutkin ordered Linenger into their *Soyuz* to prepare for evacuation. Tsibliyev jetted between his controls and the window, hoping to catch sight of the *Progress* before it crashed into Mir. He couldn't see it against the black. His screen remained a blizzard. He bit his lip and began furiously cranking the joystick that dictated the *Progress*'s flight path and, in turn, the fate of his crew.

The *Progress* missed colliding with the station by only two hundred meters—a whisper by galactic standards—like a torpedo diving away from a dead-in-the-water sub.

Facing down death twice did Linenger in. He stopped talking to the ground. He also refused to take part in his weekly medical conference. He believed that Americans should never again come to Mir. He wanted the exchange program to stop with him. Otherwise, he argued, the next man in line had a real chance of not coming back.

The next man in line was Mike Foale.

. . .

At first, Foale's arrival breathed new life into Mir. He was easygoing and good with Russian, eager to learn and friendly. Tsibliyev and Lazutkin, having never connected with Linenger, were happy for Foale's warm company. Although they continued to battle coolant leaks, the three men were something like content.

April gave way to May, May to June. Then came nervous news from the ground: they would like to try docking the *Progress* manually again.

The timing was bad. Tsibliyev was tired after diving face-first into a bubble of antifreeze, which left him feeling poisoned. He had also been conducting a series of sleep experiments, during which he had to wake up throughout the night and draw blood from himself.

To make matters worse, the Russians had decided that rather than risk shorting out the monitor again, they would stop the flow of telemetry data, which was what had interfered with the broadcast signal during the first attempt. Trouble was, that data gave the commander the speed and range of the incoming ship. Without it, he would have to estimate the approach using a handheld laser range finder and a stopwatch.

It was a formula for disaster: ethylene glycol plus fatigue plus a big rocket heading toward a last-legs space station, at a speed and range that was, at best, a best guess.

Then the brakes didn't work.

The *Progress* suddenly appeared through Mir's window too soon, and far too late for any evasive action. Lazutkin, who had been on the lookout, had time only to close his eyes and turn his head.

Tsibliyev saw the look on Lazutkin's face and knew. "Oh, hell," he said.

There was a shudder, like an earthquake. The *Progress* drove a hole into a solar array, wedged itself against the station, broke free, and came in for a second run. This time it hit Spektr, and this time it punched through the hull. The master alarm sounded. "We have decompression!" Tsibliyev shouted. "Hell, Sasha. That's it!"

Lazutkin swam to Spektr and heard a sound no astronaut had ever lived to describe: the angry hiss that air makes when it's rushing out into space.

The more immediate problem was that he couldn't close Spektr's hatch, which was blocked by ventilation tubes and no fewer than eighteen cables.

With the pressure inside the station dropping rapidly—780 . . . 700 . . . 690 . . . 680 . . . 675 . . . 670—Lazutkin began frantically pulling apart the cables. He got through fifteen of them in three minutes. The three that were left had no visible disconnect. He found a knife and cut through the first two, which turned out to be data lines. He sliced into the last cable and sparks shot up his arm. It was a power cable, and it couldn't be cut.

Tsibliyev opened up some spare oxygen tanks, hoping to buy the crew some time. Foale worked to prepare their *Soyuz* for evacuation. Wondering where in the hell the Russians were, he kicked back toward Spektr and found Lazutkin trying to find the power cord's plug. Finally he did and set about closing the hatch, but even with Foale's help, he couldn't. They had to pull it toward them, but the force of the air rushing through the hatch and out of the hole in the hull was too great. In order to close the hatch, they'd have to wait for the pressure to equalize. Too bad their blood would be boiling by then.

Struck by panicked inspiration, Foale and Lazutkin rushed to find the lid that had covered the hatch when the module was delivered. Somehow, under all of that junk, they did, and they slammed it into place.

The rest of Mir and their lives were saved. But Spektr was lost. Foale's sleeping compartment, his personal effects, and half of the American experiments were on the other side of that lid. So was pure, open space.

They had also lost the power generated by Spektr's four solar arrays, leaving the station alive but limping, lost in free drift. The crew stopped the slow roll by firing their *Soyuz* capsule's thrusters, and they managed to restore some of their power supply, but over the coming weeks, they would have to set their minds toward a "space walk" inside the station, trying to patch Spektr's hull and reconnect the cables that had run through the open hatch.

Normally the two Russians would have done it, but buried under post-collision stress, Tsibliyev began suffering an irregular heartbeat. He was scrubbed from the fix-it mission and replaced by Foale, a move that pushed Tsibliyev closer to psychological collapse. He would spend the coming days in tears. Pools of water collected around his eyes, refusing to run down his cheeks.

In the end, Star City decided to postpone the work, sending up a fresh crew to make the repair. Commander Anatoli Solovyov and flight engineer Pavel Vinogradov joined Foale in August. They suited up and headed into Spektr.

They reconnected some of the cables, and they retrieved Foale's personal effects—photographs of his wife and children and his toothbrush, all of the essentials—but they failed to find the hole that the *Progress* had knocked into Spektr. The hatch would have to remain closed, the module never again part of the machine, except in a series of photographs that were beamed down to earth, showing a battered ship that had come so close to becoming a coffin.

. . .

In light of those images and the horror they invoked, debate raged in NASA's corridors about the future of the Mir program. Jerry Linenger, who would soon become the third of the American Mir astronauts to quit the corps, railed especially hard against a continued presence on the crippled station. The sticky thing was, had the Americans decided against stamping their tickets to Mir, the Russians would almost certainly have pulled out of the International Space Station, which still had strong backing from the Clinton White House. The two remaining missions would have to be completed, if only for appearances.

Not surprisingly, however, NASA couldn't smoke out many volunteers to fill the final two slots. The sixth American on Mir, Dave Wolf, was a big-drinking daredevil who had nearly had his pilot's license revoked for buzzing houses near Houston. Wolf saw in Mir his last chance for redemption, and he made the most of it. Although he spent nearly half of his mission, between September 1997 and January 1998, wiping up water puddles and condensation, his easy spirit made him the unlikely savior of a long-money program.

Australian-born Andy Thomas, despite his limited Russian, was chosen to make the final trip. His 130-day stay would prove NASA's third-longest, after Lucid's and Foale's. Among the Russians who shared his company was a smiling, stocky cosmonaut named Nikolai Budarin.

. . .

A husband and the father of two boys, Budarin had begun his career in 1976 as an engineer for ENERGIA, Russia's massive space technology contractor. After spending more than a decade designing and building space stations and rockets, Budarin, like a sports writer who's grown tired of watching other people play the game, had decided that he'd like to try his own hand at living in one of his creations. Between 1989 and 1991, he attended classes at Star City's Gagarin Cosmonaut Training Center and, after passing the state examination, was qualified as a test cosmonaut. It took two more years of training before he was qualified to fly *Soyuz*. He also prepared for a visit to Mir.

His first stay was a relatively short one, from June until September 1995, but long enough for him to complete three successful space walks and enjoy a luxurious flight on the space shuttle, the first time a Russian cosmonaut had hitched a ride on it to Mir. (After Budarin was safely delivered by *Atlantis*, Norman Thagard took his seat on the return to earth.) His second stay, the one that he shared with Andy Thomas, was even more notable. He spent the first seven months of 1998 in space and, during that time, completed six space walks, helping to repair Spektr's damaged solar panels and earning high enough praise from the ground to become a Hero of Russia.

He had also become a favorite of the Americans. In the halls of Star City, he was quiet and serene, if a little serious; he liked to fish and to ski, and when he was asked to name his favorite hobby, "picking mushrooms" was his usual answer. Happily, a man who can get along picking mushrooms had the perfect temperament for coping with the monotony of living in space. Whenever he was in orbit, Budarin was quick to laugh, the proverbial teddy bear, warm and gentle. It took a lot to ignite his temper, even when he was forced to spend most of his days on Mir repairing the tender machines that he had helped to build. He was the sort of man who whistled when he worked.

As a result, despite his poor command of English—he spoke it like Tarzan, mostly in two- or three-word sentences composed entirely of nouns and verbs—Budarin made for pleasant company.

Charles Precourt, the pilot of Budarin's first shuttle mission, recalled a preflight drive from Houston to Galveston, Texas, when the two men spent forty-five minutes passing a Russian-English dictionary between them. Muddling through a halting conversation, he and Budarin still forged something like a friendship on their way to the Gulf Coast. During that otherwise unremarkable drive, over causeways and through swamps, Budarin, especially, discovered the joys of communicating by means other than words. Contrary to what he had been taught as a child growing up in stone-faced Russia, he learned the power of a smile. He came to understand that so long as he said whatever he was trying to say with a light in his eyes, he had no fear of his message being lost in translation.

He and Andy Thomas (whose Russian was only slightly better than Budarin's English) didn't often "speak" to each other during their months together in space. But along with Mir's amiable commander, Talgat Musabayev, the two men made a habit of eating dinner together and taking comfort in each other's company, even if it was only in a calming silence. When something really needed to be said, wild gesturing became the official language of Mir, usually punctuated with laughter after the men realized how ridiculous they sometimes looked.

Fortunately, because they both spoke German, Musabayev and Thomas were able to have real conversations, most often away from Budarin, so that he wouldn't feel left out. Over Thomas's litany of scientific experiments, they would tell each other war stories, and, in time, they found plenty of common ground talking about music and art and what they missed about home. Those bonds proved important over the course of the coming weeks, when Mir struggled through a few more of its mishaps, including sweltering temperatures and a small, contained fire that pushed carbon dioxide levels dangerously high. Almost laughably, the first of Budarin's and Musabayev's planned space walks to repair Spektr was scrubbed when they couldn't open the airlock's hatch, one problem compounding another.

Budarin had bent or broken three wrenches trying to crack the hatch's code before he gave up. That kind of persistence, coupled with his surprising lightheartedness throughout the episode—after heading back inside, he had looked at Thomas with a smile and a shrug as if to say, "Shit happens"—was what most impressed the Australian about him. Here was this man who boasted a Russian's stoic calm coupled with the backslapping familiarity of an American. It was a rare combination, especially for a cosmonaut, and especially on the troublesome Mir. Nikolai Budarin was, in a lot of ways, the perfect space sidekick. He was the best of both worlds.

. . .

At first, the International Space Station appeared to be built out of the worst of them. The newly forged relationship between Russia and the United States got off to a rocky start when the Americans put pressure on their new colleagues to de-orbit Mir. Trying to keep it aloft, well past its expected drop-dead date, was proving a distraction in getting the new station off the ground. But for the Russians, losing Mir was like losing a limb. There would be a phantom itch when it was gone.

Though they were cash-strapped and knew, in their sheltered hearts, that it was time to say goodbye, they resisted letting go until December 1999. The announcement was made with the solemnity of a eulogy. Deaf to that farewell, a new consortium, MirCorp, tried to find a way to rescue the station with private money. There were also last-ditch political fights to save it, with the Communists desperate to keep this last great Red Star in the sky.

But inevitably, an unmanned *Progress* was dispatched to Mir in the winter of 2001, more than fifteen years after the station's first module had been launched. The *Progress* docked, and like a tugboat, its rockets were used to push Mir closer and closer to the earth's unwelcoming atmosphere. Its time in orbit finally ended on March 23, 2001, when the station lit up like a funeral pyre before its surviving fragments splashed down into the South Pacific. There were tears in Moscow and among the 107 men and women who had lost one of their more memorable homes.

Perhaps it's not surprising, then, that having been born during Mir's controversial demise, the International Space Station sometimes seemed a bad seed.

Almost from the beginning, the Russians had fallen behind on their funding commitments and, more important, on their module construction. Although Mir had given them the inspiration for a host of design improvements—they sought to add lights and sensors to aid in docking, provide quick disconnects for the cables that ran through open hatches, and reroute cooling lines and electrical cables to prevent moisture buildups and leaks—they seemed to lack the will to turn their lessons into hardware.

Zarya, the station's first building block, was launched late, in November 1998, not long after Budarin had returned from Mir, as though in some cosmic way he had become the weight on a pulley. Two weeks later, the shuttle *Endeavour* launched with the first American module—tiny, coral-colored Unity—on board. Ground control tried to plaster over the cracks that had formed in the new partnership, pouring out good feeling while watching the shuttle climb into the sky: "We have booster ignition and liftoff! The space shuttle *Endeavour* with the first American element of the International Space Station, uniting our efforts in space to achieve our common goals."

The shuttle's crew captured the still-unmanned Zarya with the Canadarm and, with relative ease, brought it together with Unity; the two modules were permanently connected over the course of three space walks. It was a birthing-room moment. At last, station was the object of hope and not just hand-wringing. Like Forrest Gump, omnipresent Russian Sergei Krikalev—there he is again!— and American astronaut Bob Cabana were the first men on board, swimming together through the hatch that had been opened between the two modules, a symbolic shared step over the threshold. Their kicks provided a much-needed injection of optimism to the project, all but forgotten when station entered its troublesome middle life.

By February 2000, more than a year had passed since Zarya and Unity had been brought together. But they had remained empty, because construction delays in the station's third and most critical component—Zvezda, with its propulsion and life support systems— had left it uninhabitable. (Not coincidentally, the module was the first that the Russians were to finance on their own.) Worse, Zvezda's absence had left the embryonic station unable to keep itself in orbit; the two modules had lost nearly a mile in altitude each week. Twice shuttles had to be sent up to push them into a higher orbit.

The irony was that the Russians had been brought on board to speed up construction, to tap their experience, to make things go more smoothly. The collapse of the Soviet Union and subsequent economic chaos conspired only to slow it down, adding billions to the cost of building the International Space Station and leaving its foundation on shaky ground. Rather than proving the largest construction project since the Great Pyramids, it was starting to look more like the failed Tacoma Narrows Bridge, twisting in the wind.

Finally, however, in July 2000, Zvezda was launched and locked into place. Hopes for a smoother, brighter future were almost immediately scuttled when the first human elements were about to be introduced to station: the veteran Russian cosmonaut Anatoli Solovyov quit Expedition One when American Bill Shepherd was named commander over him. Fortunately, the trio that eventually became the first to call the International Space Station home—Shepherd, Yuri Gidzenko, and Krikalev (one more time!)—helped heal the program's early bruises.

On November 2, 2000, they kicked off what would be a continuous and surprisingly harmonious manned presence in space, remarkable not only for its endless scope but also for the composition of the crews that made it possible. Until station became large enough to host more than three visitors at a time, the plan called for each expedition to consist of two Russians and an American, followed by two Americans and a Russian, until the two countries had sent enough of their pilots and scientists together into space for them to blur into one great string of names and faces, the citizen soldiers of a new country. By the time Expedition Six arrived on its doorstep, station had already hosted fifteen men and women who together had lived for more than two years on board. The stories and memories of their work and play had eclipsed the early fights and troubles. They had taken what had looked destined to become a battleground and made it into a shelter. They had taken what had been given to them and built it into something larger.

. . .

Expedition Six, having just started to emerge from their private grieving chambers in the days following *Columbia*'s loss, would see the International Space Station become more for them than it had been for anybody else. Ken Bowersox, Nikolai Budarin, and Don Pettit would see it as more than a brightly lit place to grow protein crystals and drink their coffee through straws and run on a zero-gravity treadmill. Instead, they would come to see in it a comfort, the makings of a sanctuary. For them, station would become a home, and they would soon become a family built on love and trust and experience, like the one each man had left behind on earth. This new family would be captured in photographs and written about in letters. And somewhere along the way, having been battered by so much time and distance, the lines between Russian and American, man and machine, and even earth and space would begin to disappear. They would break up and vanish, like that finger of white smoke over Texas.

5 GONE

Every so often, one of their orbits followed *Columbia*'s last flight, right over Houston. For days beforehand, Ken Bowersox's three boys would track station on their computers, and because they were old enough to understand a little of what he knew about the universe, they could calculate almost to the minute when they might catch a glimpse of their dad's second home. The timing had to be just right—it had to be dark outside, but the night had to be young enough for the sun to have dropped just below the horizon, still reflecting its rays off the space station's solar panels. There couldn't be a cloud in the night sky, and it couldn't have been so hot for the city's haze to have stayed draped over Clear Lake. Only once or twice in a very long while did everything fall into place. On those perfect nights, the boys gathered on their front lawn, their feet in the cool of the grass, and strained their necks until they spotted a small, steady white light coming up over the trees. They followed that light for as long as it took to cross their starlit sky on a smooth, predetermined path, the same path that would carry it over South Africa by the time they went to bed.

Six-year-old Luke, Bowersox's youngest son, didn't quite have a handle on that part of the deal. Speed and time and distance are relative things to a kid. During each of the long days he waited for station to pass overhead, he made up his mind that this would be the night he would catch his dad and bring him back down. As soon as the light came over the trees, he'd begin chasing it, taking off down the street, hoping to cover enough ground, enough of the curvature of the earth, to earn even one more second in his dad's line of sight. And always, the light disappeared.

. . .

On the starry night she met Don Pettit, Micki Racheff, a deejay with a taste for eclectic music and social engagements, was recovering from a hangover and had just begun work on a new one. She had gone from her home in Santa Fe to a house party outside of Los Alamos only because her friend wanted some company while she hunted for a new man. (The party was well-stocked with Beakers from the laboratory; for a woman with a ticking biological clock, it was the sort of gathering that almost guaranteed her imaginary offspring would sport giant frontal lobes.) Wandering into the kitchen—hoping to find just something to begin healing herself with, not a future husband—Micki spied her friend enjoying an animated conversation with a happy man and joined in. Pettit was in the middle of recalling his recent research on board the "Vomit Comet," admitting that he had, in fact, been sick all over himself. But now having seen this pretty woman with dark eyes sidle up— and seeing, too, that she wasn't in the mood for puke stories—he switched gears, explaining how the bubbles in her freshly poured glass of champagne would dance in zero gravity. The rest of their conversation that night was just as romantically dorky. Micki left thinking that this guy in the kitchen was funny and obviously really, really smart, but it wasn't until they met again a couple of days later, alcohol-free, that she decided he was cute, too, smiling and stammering through his endearing brand of breathlessness.

She also liked that they were both Westerners (Micki was from Wyoming) because she believed that the Continental Divide ran through relationships, too. She had always felt that people who had seen mountains were different from people who had not. They were more limitless somehow.

Their marriage proved it. In August 1995, they made the long journey to Australia—opting, true to form, to visit its barren west rather than its urban east—and decided that it was a fine place to get hitched. Staying at a sheep ranch that took in lodgers, they mentioned to their hosts that they planned on getting married. The ranchers first explained the legalities of union Down Under; next they volunteered that their daughter had recently been married at the ranch, and they'd happily put out the same spread again. After Don and Micki drove up the coast for a spell, stopping in Monkey Mia to play with dolphins in the warm water, they returned to an unforgettable setup. About a dozen other leather-faced ranchers had been invited and come on in from the outback; Don, who had planned on wearing shorts, was scrubbed clean and dressed up (and squeezed into polished shoes two sizes too small); the top half of the daughter's wedding cake had been taken out of the freezer and set on a table. Under perfect skies, in the folds of rugged, almost breath-catching country, Micki Racheff became Micki Pettit, by witness of sheep and tearful strangers.

. . .

She didn't know it then, but Micki had signed on for a life as uncommon as her nuptials.

A couple of years after they had started dating, Don was granted his third interview to become an astronaut. Trying to feel optimistic, he flew to Houston and sat down before a stern panel. But when Micki picked him up at the airport upon his return, she could tell just by seeing him that the interview had not gone well. The candidates had each been told that they would soon receive a phone call. What they weren't told, but what they all knew, was that if they got a phone call and heard one man's voice, it was good news; if they got a phone call and heard another man's voice, it was bad. Sure enough, one afternoon Pettit picked up the phone and heard the Grim Reaper's telltale wheeze coming across the wires. Don and Micki sat quiet, heartbroken, but together. And together, they began to accept that Don would probably never make it into space. They decided to move on with their earthbound lives, Micki on the radio, Don at the lab.

Then, suddenly, just a few weeks after they came home from Australia, Don was granted another interview, his fourth. It felt like his last, best shot at his dream, and, already like astronauts, he and

Micki prepared down to the smallest detail for the big event. They fought to make sure that they didn't overlook whatever tiny thing had been holding him back, whatever it was that made him good enough to earn interviews but not a seat on the shuttle. For maybe the first time in his forty years, Don bought himself a good suit and tried his best to put himself together. Micki, looking in the mirror with him, made sure that even his socks would pass muster. Immaculate, Don boarded a plane for Houston, and Micki waited back in New Mexico for his return, never allowing herself to imagine that those anxious nights alone might make for good practice.

It was months before the phone rang again. When it finally did, in April 1996, it was someone other than the Grim Reaper calling. It was Don, calling from that cramped cottage in storm-lashed New Zealand, and here he was, telling Micki that he was about to have the chance to travel much farther away from home. They screamed and cried and laughed at each other over the Pacific. It was an unreal moment. That cute guy in the kitchen was now a full-fledged astronaut, and he and his wife would have until August to quit their current lives and head for Houston.

At the radio station, Micki announced to her friends that she would be leaving. With old images of the smoothed-over Apollo wives and their permanent smiles springing to mind, one of Micki's friends joked that she had better pick up a pillbox hat.

Later, driving in their short convoy to Houston—Don in his junky pickup truck and Micki in her sedan—they filled the lonely hours in West Texas by talking on their CB radio, having assumed truckers' handles for the trip. First Micki Racheff had become Micki Pettit. Now she had taken to calling herself Madam Pillbox.

But it was months before she really took the change to heart—not until Don was dipped into the Johnson Space Center's neutral buoyancy pool, the massive tub in which apprentice astronauts enjoy their first chance to splash around in spacesuits. Micki went to watch (wearing a security pass that read ASTRONAUT DEPENDANT) and to take pictures. Don was below her feet, at the bottom of the world's largest swimming pool, trying to regulate his breathing, weightless for the first time since he'd been sick on a plane that was falling out of the sky. That was when Micki first realized that this new life of hers was real, that she wasn't just floating through some elaborate fantasy, a dream, or her husband's sometimes too-fertile imagination. Suddenly she was down there at the bottom of the pool right alongside Don, trying to keep her own breathing under control. Though she kept the thought to herself, she couldn't help thinking: Oh, shit.

She was finally an astronaut's wife.

. . .

The role has changed since that night when Mrs. Armstrong, Mrs. Aldrin, and Mrs. Collins wore red, white, and blue. Then, an astronaut's wife had her background and credentials as closely scrutinized as her husband's had been. NASA didn't want any of the women saying or doing anything even remotely untoward, anything that might cause the American public to withhold a single ounce of the love and energy that had been pouring into the program. That meant that most of the wives were cut from the same (spotlessly clean, neatly arranged) cloth: they were pretty and deferential, doting mothers, and uncomplaining homemakers. ("You worry about the custard, and I'll worry about the flying," Frank Borman, the commander of *Apollo 8*, had famously said to his wife, Susan.) Most of all, they were to make for good television. During launches, their lawns would be covered by reporters and satellite trucks; in between, they would express their pride in their husbands in feature reports and on the pages of *Life* magazine. They were to glow through all of it. The wives were an integral part of a giant publicity machine, the women behind the men destined to become heroes.

It was an uneasy life in a lot of respects. Their husbands were often absent and, at best, part-time fathers. Stress was a permanent fixture in their lives, most acute when their husbands were on their way to space, in space, or on their way back from space. (Most of their homes were equipped with "squawk boxes," which relayed the chatter between the rockets and Houston, but someone from the office was usually assigned to listen along with them, so that the transmission could be disconnected in the event of trouble.) These brave women coped with fear, infidelity, loneliness, and their own pressures of performance.

Not surprisingly, the combination took its toll. Some of the wives, including Susan Borman, began drinking heavily. One of them, Pat White—the widow left behind by Ed White, killed in the *Apollo 1* fire—committed suicide many years after the accident.

But for the most part, the wives were exactly what they were expected to be. They were military wives, and their children were military children. They were all too accustomed to their husbands and fathers leaving for long deployments and finding themselves in mortal danger. The families, in turn, assumed a stoicism that wouldn't have seemed out of place on the hard road to migrant California, passing by Tom Joad's jalopy. They prepared themselves for doom, having learned to assume that one day, two men in crisp uniforms would knock on their door with white-gloved fists and tell them that there had been an accident. Even if they were lucky enough to duck firsthand grief, no doubt they had been brushed by it. Someone they knew had lost someone close to them in a fire or a wreck or a dogfight. They had probably attended the funerals, and they had probably brought over a hot chicken dinner for grieving widows and orphans, and they almost certainly had seen a flag lifted from a coffin, folded into a triangle, and handed to a stone-faced woman dressed in black.

Because almost all of the pilots who become astronauts today are still plucked from the military, their wives, too, remain what they always have been. Most of them are not the snow-white trophies of old, and most of them don't turn a blind eye to the girls in other ports, the way they once did. But some part of them still exudes the air of widows-in-training. They are the sort of women who have grown into their hard shells. They are also private, formal, careful, pious, and sacrificial. Annie Bowersox, Ken's wife, is built true to the prototype. She knew what she had signed on for, and she knew what was expected of her. On those rare occasions when Mr. and Mrs. Bowersox attended astronaut socials, she understood perfectly what was meant when women were asked to wear "church dresses." She had already filled her closet with them.

But when the class of 1996—forty-four members strong, a cull so large that they were nicknamed "the sardines" because there wasn't enough office space to fit them all—and their families received their first such invitation, Micki Pettit was baffled. She had no idea what a "church dress" was. It wasn't until after some whispered consultation with a few of the other wives that she discovered, to her mock horror, that a church dress was one that made a woman look like a barrel. Curves are frowned upon, and cleavage is strictly verboten. The astronaut business is a serious one, she was told—all the more so, ironically, when it is conducted on the ground. There are cliques and favorites and rituals and rites of passage, and with the sardines especially, there was a long line ahead of them to get into orbit, fraught with missteps and peril. No one wanted to stand out, at least not for the wrong reasons, and that included their wives showing too much tit.

Trouble was, no matter what Micki had on, her husband couldn't help standing out, ignoring even his legendary classroom pronouncements on the color of rocket fuel. Since Skylab, when astronaut-scientists first began joining towheaded military fliers on missions into space, the "civilians" have been looked at as oddities and interlopers, as though they never quite fit with the program. They were cargo, and worse, they spent their precious time in space growing tiny plants and blowing bubbles (which, to NASA's navy and air force men, was a little like obsessing over how to make perfect toast while riding the world's greatest roller coaster). Spaceflights suddenly felt like high-school cafeterias, with the jocks and the geeks staring at one another from separate tables across the room.

It didn't take long for there to emerge a further divide, this one among the scientists themselves. First, there was the majority, all of those mission specialists who dreamed of becoming the envy of their gravity-bound peers, zooming into space on the shuttle; conducting simple, camera-friendly experiments; and returning home in two weeks with a lifetime of stories to tell. Then there were those very few scientists who wanted to land themselves on station, out of reach of ground control and its rigid demands. They yearned to be cut loose, free to explore each and every idea that filled up their dreams. The shuttleheads saw something bizarre in those fantasies, something lonely and rudderless. (They also didn't like the idea of spending years training in Russia.) But from his first weeks in Houston, Don Pettit set his sights on measuring his time in orbit in months, not days. He made it plain that he didn't mind going it alone, and he didn't mind one bit if he was sent up there and forgotten. He had been saddled with the image of the outcast for so long, it didn't even occur to him to fight it anymore.

He might have even gone out of his way to cultivate it. Because of the limited number of houses for sale near the Johnson Space Center that summer, and because there was such a large incoming class, most of the recruits had already toured one another's homes. As soon as one of them made a down payment, the rest of them nodded, able to remember from jammed open houses its layout and lawn ornaments. The Pettits had gone in for one of the bigger homes, partly because they were thinking about starting a family, but mostly because Don wanted a three-car garage, which he promptly filled with his tools, experiments, old electronics, and an entire jet engine. In addition to its square footage, their home also boasted a gas fireplace, which, because it's usually plenty hot in Houston, had found a place in the memory banks of each of the families who had seen it. It seemed to most of them like a loopy extravagance.

It also had fake logs stuffed into it, which Don couldn't abide: if he was going to watch something burn, he might as well watch something interesting burn. And so he set about replacing the logs with a diorama of a miniature village, complete with scorched rooftops and panicked residents jumping out of their windows. Whenever he flicked on the gas, the town would appear to go up in smoke—and so, too, did another wisp of his reputation each time a joyless visitor asked to see his latest creation.

. . .

Fortunately there were plenty of misfits within the astronaut corps—not only fellow civilian scientists but also the international

astronauts who joined NASA, a little more arty, a little more experimental than some of their American counterparts. They were a little more out there, and Don and Micki joined them on the edges of
the fraternity. In addition to Chris Hadfield, the Canadian guitarist,
the English scientist Piers Sellers and his wife, Mandy, became good
friends. So, too, did Ilan Ramon, the Israeli payload specialist, and
his wife, Rona. Unlike some of NASA's military fliers, they would
meet at places other than church. They would get together for dinner and drinks, for long nights of music and debate. Those nights
made Houston feel more like home and getting into space feel less
like a war of attrition.

The Pettits were soon distracted by bigger battles, anyway.
Shortly after moving to Texas, they had tried starting a family. It
was not easy. After a couple of futile years, there came long, painful
rounds of tests and injections and the tough questions that childless
couples have to ask themselves. There was more failure and heart-
ache until Micki finally became pregnant—Don was in Russia
when she found out, and she had to share the good news over the
phone—and gave birth to twin boys, Evan and Garrett, in November 2000.

At last, they were complete . . . Until Don Thomas was found to
have absorbed too much radiation, and Don Pettit—the outsider
who had finally begun working his way in—had just three months
to say goodbye to the home life he had so long wanted and exchange it for another.

• • •

On the grounds of the Kennedy Space Center, past the swamps and
ragged shoreline made foul by red tides, there is a house on the
ocean near Cocoa Beach. When its shutters were first opened, it was
designated a party palace, a homespun gin joint in which hell-
raising astronauts could kick back. During those early, ribald days,
kicking back meant keggers that lasted till dawn, bonfires built high
enough to light up the night sky, ashtrays filled to overflowing, and
wives (and those girlfriends the wives did or didn't know about) being led by their hands to the bedrooms downstairs. It was the literal

last resort, a sort of hedonistic Eden for all of those Adams who figured that their cradle on a rocket was as good as a deathbed. In between assignment and liftoff, whenever they weren't in training or put on public display, they made sure to live life as hard and as well as they could. They were like college students whose summer was coming to a close, only with the risk that they might never again see winter turn into another spring.

Today that feeling still hangs over the place, but it's reserved for gentler goodbyes. Crew members are bunked in private quarters, under semipermeable quarantine, for weeks before launch. But in the waiting, they are allowed to escape and come here to spend some time with their wives and husbands (in the case of STS-113, of course, there were only wives). Catered dinners are delivered, and fourteen distracted people sit down in the dining hall and, if only for as long as it takes to finish their dessert, they make small talk, trying to avoid thinking about what's coming down the chute. After the last of the food is polished off, the group breaks up into couples. Some sit on the balcony, watching the sea roll in. Others might still head for those bedrooms downstairs. Don and Micki decided to take a long walk on the beach, to find a little fold just for themselves in the dunes.

They had sweaters and jackets on to ward off November's chill. Through the dark of night, they could still see the white foam left behind by the waves running up the beach. The light from the moon and the stars helped, shining off the water. Even when a thin fog started creeping in, the sky stayed clear. It was a night built for bad love songs, but really, it was just one of those beautiful nights.

Hand in hand, they talked. They had already made formal plans, just in case: like Ken and Annie Bowersox, they had made sure that Don's will was in order, that their insurance was paid up, that Micki and the boys would be looked after if things went very wrong. But on this night, they talked about what Don's leaving forever might really mean—not on paper or in bank accounts but in clothes still hanging in the closet and shoes gathering dust by the door. They talked about what Micki would do, where she would live, what she would keep of Don and what she would have to throw away. They talked about the boys and what Don wished for them. They talked about how, no matter what happened, they were glad to have met that night in a kitchen outside of Los Alamos. Their lives had been shaped by that moment, that hangover, just as they knew that their lives might be shaped this time by an O-ring or a loose ceramic tile. They left nothing held in reserve. They kept no secrets. They said every last thing that needed to be said, and then they turned around, their feet pushing into the white sand, and together they were led back to the beach house by the lights and the noise. Finally prepared that their marriage might continue from that moment on only in memories, they hugged and kissed and said goodbye, their breath turning solid and joining the fog, lifting into that beautiful night.

On the ride back to her nearby hotel, Micki sat next to Robin Wetherbee, the trembling wife of Commander Jim Wetherbee, heading into space for the sixth time, one short of the American record. Micki had been watching her all night, thinking about how anxious she seemed, this veteran of farewells—while other, younger couples, including the Bowersoxes, had breezed through their goodbyes by treating it like any other, as though their loved ones were leaving for a weekend conference in Sacramento rather than blasting off into orbit.

"Doesn't it get any easier?" Micki asked finally.

"No," Robin said after a long while. "It only gets worse."

. . .

Whatever internal drama *Endeavour*'s crew went through during their on-again, off-again launch schedule, their wives went through it sevenfold. Since *Challenger*, there is a private rooftop, several miles from the launchpad, from which the families watch the shuttle lift off. If, seconds later, the sky is lit up by an explosion, the isolation ensures that shattered husbands, wives, and children won't have their hysterical blindness worsened by flashbulbs.

In the hours before—alone or with their kids—they gathered their nerves in their cheap, frayed hotel rooms in Cocoa Beach, until there was a knock on the door. Then they headed downstairs and boarded a bus that took them to their rooftop, dropping them off into their own private worry pit. There they waited, watching a clock that counted down to ignition. Sometimes it seemed to take forever for it to reach zero.

Micki had brought her toddler twins to Florida with her. They were each a handful, and caring for them had helped distract her during the waiting for launch. The first time Don and the rest of *Endeavour's* crew had climbed through the hatch—the time that saw them climb out again ten minutes later, thanks to that ill-timed oxygen leak—she learned of the scrub while she was still getting the boys ready to head outside.

Together, along with their husbands, *Endeavour's* wives and children returned to Houston after the Canadarm was bruised; the eight-day respite that their men enjoyed only prolonged their own agony. That first single night together had felt like a gift, a second chance to make certain that everything that needed to be said had been said, but after that, it had been impossible for them to live out their usual lives, to get through their usual routines without their stomachs clenching or lumps rising in their throats. No matter how busy they had tried to keep themselves, they had noticed the emptiness in their homes. Even Annie Bowersox, who had been so calm in the face of the stress of saying her goodbye, had moments when her heart stopped in the quiet. Her husband had defied the odds so many times already, on aircraft carriers and above the desert flats and four times in these beautiful rockets. And like Robin Wetherbee, some small part of Annie wondered whether his good luck was destined to run out. She had also decided that she would have felt better if her man was in the driver's seat. He had always done everything he could do to bring himself home. This time, if he didn't come back through their front door, she didn't much like the idea of having somebody else to blame.

But like their husbands, the wives had their destinies taken out of their hands. Back in Florida, they had just boarded the bus when liftoff was canceled again, this time because of bad weather over those two podunk towns in Spain. They began to feel like racehorses who had been led out of the paddock and into their gates, left keyed up and waiting for the starter's pistol that never seemed to fire. They returned to their hotel rooms, and they tried to get some shut-eye, but most of them didn't. Most of them, even though their hopes had been carried only partway up, found it impossible to come down enough to sleep. Those who did woke up with starts and in cold sweats. Daylight couldn't come fast enough.

. . .

Mercifully, they made it to the rooftop. Annie Bowersox breathed easily, her children by her side. Robin Wetherbee trembled, blaming the cold. Micki Pettit held Evan in her arms; someone else held Garrett. She pointed to the big rocket in front of them. "Daddy's in there," she said. All of them could see the shuttle, an almost impossibly bright white under such hot lights, and they could see the clock, ticking down. With their gaze bouncing between the shuttle and the clock, as though they were watching a tennis match, they waited for a hitch that never came.

With nine minutes left on it, the clock stopped on its temporary hold, but after a few minutes it started up again. Terminal count had come and gone. All systems were go.

With seven minutes remaining, they could see the White Room pulling away from the shuttle. Now their husbands were locked in good and tight. Maybe, someone joked, this stupid thing was going to lift off after all. Maybe, someone else answered, and they shared a nervous laugh.

When their clock had ticked down past three minutes, the wives could hear the main engines gimbeling. If they squinted, they could see the shuttle swinging against the tower, bucking like a bull waiting to be cut loose.

The next minute was just long enough for Micki to decide she needed to throw up.

The clock ticked down to ten, nine, eight . . .

And finally, down to zero.

Jets of orange fire were spit out of the back of the shuttle— mostly out of the solid rocket boosters that had been strapped to the sides of the external tank, but also out of the shuttle's three main engines. Five bursts of fire became one, corralled by the trenches that had been dug for it. It ran into the walls of water that had been released along with it, creating two thick, enormous clouds of steam on either side of the launchpad, almost framing the shuttle in that instant when it lifted free of the tower.

It took a few seconds for the sound to carry as far as the sight. It rumbled in low and next turned into a roar, a sound that could be felt more than it could be heard, first in the feet and then in the stomach and then in the throat.

The shuttle continued its long, loud climb. It had left in its wake a spreading cloud on the ground, reaching its fingers out into the dark, and a fat vapor trail, lit up by the fire that still blew out of its ass.

On their rooftop, the wives watched with their hands over their mouths and tears in their eyes. Annie Bowersox looked as though she was staring at the flag while the anthem played. Robin Wetherbee bit into her bottom lip. Micki Pettit, whispering under her breath so that Evan, still in her arms, couldn't hear, began pulling profanely for the shuttle to keep lifting into the night sky.

"Go, you fucker," she said. "Go, go, go."

The fucker obliged.

And in that moment, just when it seemed as though their men would make it safely into space after all, something switched over in the wives. Suddenly the spectacle outweighed the sensation, and their worry was replaced with a kind of wonder. By the time the shuttle had traveled twenty-seven miles straight up, twisting ever so gently on its way to finding its orbit, and its solid rocket boosters had been jettisoned on the heels of a telltale flare, the wives gasped not out of fear but from awe. They were no longer seven women standing on a rooftop watching their husbands ride fire. Instead, they had joined the tens of thousands collected along Florida's coast that big starry night, standing on hotel balconies and beaches and the gravel by the side of the road, feeling subsumed, insignificant, as though there was no greater cause than to follow this beautiful light with their wet eyes on its way to the end of the earth.

Along with their husbands, they had made the turn.

. . .

They watched the rest of the journey via satellite, packed around monitors that had been set up on their rooftop. After *Endeavour*'s crew had become weightless, having cut the last of their strings, their wives did, too. An incredible burden had been lifted from them, and now they hugged, relieved and happy but also wrung out and left unsteady by their own journeys. They climbed back on board their bus, quiet and coming down, were dropped off at their hotels, and fell into bed, ready to wake up early the next morning and catch their flights back to Houston, ready to return to the new version of their old lives, however incomplete.

. . .

Unlike their husbands, it was a long while before they were able to settle into anything that resembled a routine. First they came home to Thanksgiving, to preparing dinners and visiting friends. Micki and the boys were guests in the home of another astronaut, Duane "Digger" Carey, and his wife, Cheryl. Like Annie Bowersox, however, Micki was distracted by the work going on at the International Space Station. The truss installation was broadcast on NASA TV, and Micki excused herself from the table to watch it. The cameras were never on Don, who was inside helping to operate the robotic arms, but occasionally she heard his voice. It was filtered through static and fatigue, but even when the tenor of it was nearly washed out, she could still tell it was him. For most of his time away, their voices were all that would carry across the divide between them.

Nikolai Budarin usually held dominion over the space station's Internet phone (Ken Bowersox was proficient in Russian, but if Budarin wanted to have a real, rapid-fire conversation in his native tongue, he was dependent on the friends and family he had left behind), but for a few minutes each week, Bowersox and Pettit could call home. Their conversations were surprisingly ordinary, dominated by updates on their growing children, snippets of front-page news, and whatever mail had arrived.

Things were more interesting on Saturdays. Shortly after the wives had returned to Houston, Annie and Micki watched technicians tramp through their living rooms, installing videoconference units. Once a week, early on Saturday morning, on a supposedly private channel, they would have between fifteen minutes and an hour to see and talk to their husbands. (Micki found out that the conversations weren't as private as she thought after she gave Don a playful flash. The following morning, she received word from the technicians that she should probably keep her bathrobe done up.) The videoconferences were especially good for Bowersox and Pettit to catch a glimpse of their kids.

Bowersox's three boys were old enough to take part in the conversations, showing him work they had done in school and opening birthday presents in front of him, but Pettit's twins didn't have a firm grasp on modern communications techniques. He had figured as much, which was one of the reasons he had brought up his didgeridoo.

On the ground, he would come home from work, and they would drag it to him, and he would pick it up out of their tiny hands and blow into it, and they would laugh at the strange noises he made. Now Micki would bring them in front of the camera, and Don would play his didgeridoo, and the noise would crash through the miles and into their ears, and the twins would look at each other and clutch each other and break into laughter, the same as they always had. In the way that Don Pettit had been reduced mostly to a voice for Micki, for his kids, he was the sound that came out of his didgeridoo. He was a low hum and a crackle on their Saturday mornings, another one of their cartoons.

. . .

By Christmas, while the twins tore open their presents and pulled tricycles out from under the tree, their father was closer to a distant memory, a vague recollection of a smiling face. Micki and Annie, too, had gone from missing their husbands to growing used to the new rhythms that had taken over in their absence. They had begun to stretch into the other side of their beds.

As often as they liked, members of NASA's family support crew would come over and look after their kids for a few hours (Micki leaned on them more than the private Annie did), and they would go out and buy the groceries or take the car into the shop or sit down on a park bench and exhale. These were restorative moments. In some ways, having their men in space was easier than having them in Russia. Then, they would see each other just often enough to make the visits feel more like disruptions than oases. For the wives, dragging their kids back and forth was draining, as was the jet lag, as was the wondering what their husbands were up to. Now, Micki and Annie knew that there was only one place their men could be, and soon enough, the messy business of real life had replaced most of their longing, filling the void that would remain open until everyone was together again, safe at home.

As they did in space, the days blurred together, turning into weeks, into months. In the meantime, Micki and Annie had become acrobats. They had become expert jugglers and plate spinners. On the surface, their routines looked nothing like the ones that their husbands had latched on to. But at their heart, they were the same. They used them to find traction in a world that had been turned upside down. Their lives had their own restraint bars; they had found their own versions of ordinary.

Almost mercifully, nothing much changed until January 2003, when Annie's and Micki's phones rang with an unexpected proposition. Because of Mike Lopez-Alegria's Iberian heritage, plans were afoot for the crew of STS-113 to tour Spain on a kind of publicity junket, talking to schoolchildren and touring air force bases. Annie and Micki were asked whether they would like to fill in for their still-absent husbands. For the first two weeks in February, they could escape, maybe spend a little time in the sun, maybe take a dip in a hotel pool or two, and by the time they were back in Houston, their men would be only a few weeks from coming home. Spain would help move along the wait.

Both women jumped at the chance. They made plans for friends to take care of their kids, and they went shopping for some new clothes. Annie and Micki had each begun packing her bags, open on her bedroom floor. And when they went to sleep on the last night in January, they dreamed of Barcelona.

. . .

When Micki woke up early the next morning, excited and optimistic, she remembered that *Columbia* was coming home. She and
her boys had watched the launch from their couch, and now she
thought it would be fun to watch what they called "the big rocket"
land. She went into the kitchen and got their milk ready, set them
up in the living room, and switched on NASA TV. Everything was
quiet, all blank faces and whispers. Micki thought she must have
had the times wrong, somehow mixed things up in the hour that di-
vides Florida from Texas, and *Columbia* had already touched down.
She flicked over to CNN, hoping that they might replay footage of
the landing. While she was waiting, she got lost in the start of her
day's routine, until she heard the announcer say—or she thought she
heard the announcer say—that Houston had lost contact with the
International Space Station.

That sat her down. She began flipping between NASA TV and
CNN, trying to piece together an incomplete story. On the former,
she could hear a disembodied voice repeating again and again, "*Columbia*, Houston, comm check," and she could hear that there was
no reply. On the latter, she watched the horrible footage come in
from Dallas of several vapor trails too many.

She called Cheryl Carey, her Thanksgiving host and wife of
Duane "Digger" Carey, who was working in ground control that
morning. Cheryl had been trying to reach him to find out what was
going on, but even before the film from Dallas had come in, she and
Micki knew in their hearts that something terrible had happened. It
was a dread that they could feel in the pits of their stomachs, the
sort of bad feeling that made their faces go hot.

"Is it gone?" Micki asked. "It's gone, isn't it?"

"Yes, Micki," Cheryl said. "It's gone."

. . .

Annie and Micki, not really knowing what to do next but needing
to do something, made the first round of phone calls, mostly to
other wives who might not yet be awake. Soon enough, their phones
began ringing all on their own, Houston's massive built-in support

network having sprung to life. But with each call, with each repetition of the same few lines of sympathy and comfort, they began to grow numb to it all. They lifted out of themselves somehow, went away, only every so often allowing their minds to come back to those new widows huddled together in Florida—"Oh my God, Lani, God, Rona, what are they going to do?"—knowing that all they could offer them, at least for now, were prayers. To make themselves feel less helpless, to take back control, they began their Saturday morning routine like robots, making breakfast for the kids (life hadn't stopped for them), picking up toys and sorting laundry, and putting on some coffee. They didn't snap out of their trances for more than an hour, when each of them heard knocks at her door.

When they opened them, there were support staff lined up on their front steps, dispatched from the Johnson Space Center, offering hugs and whatever else the women might need. Both women said that they wanted to talk to their husbands. They needed to hear their voices right away.

Micki got on the line with Don first. When their conversation started, Micki was surprised at Don's coolness. He sounded strangely calm and detached, as though he was giving a press interview to a reporter instead of talking to his wife, saying all of the platitudes that his years of training had prepared him to say, believing in all of the things that he had been taught to believe in. He said things like, "We're hopeful that their safety systems triggered, and that their parachutes deployed, and that we'll find them around campfires in the fields and hills." And Micki would wait for him to finish, nod to herself, and then say, gently, "Don, I don't think you understand," and Don would respond to her gut checks with more empty, unfounded hope. Already, in the minutes that had passed since Expedition Six had been delivered those nine awful words, he had managed to trigger his own safety systems. He had put all of that time and distance to good use, building up a wall between himself and a reality that he was now too removed from to take in. Undeterred, Micki chipped away at the divide, not because she wanted to hurt him but because she wanted him to know what she knew. She wanted him, in that moment, to come back down to earth and look up into the sky and see *Columbia* fall to pieces, so that he might come to understand that there would be no parachutes.

Finally, painfully, he did. Micki could hear the cracks opening up in Don's voice, and then she could hear his half breaths, and then she could hear his sobs. He had broken down: "But they're my friends, Willie is my friend, and Ilan is my friend, they're all my friends, every one of them, and we were just talking to them, Willie and I were playing chess, and they can't be gone, just like that, they can't be gone."

"Yes, Don," Micki said. "They're gone."

. . .

For weeks after *Columbia* was lost, Micki's and Annie's phones continued to ring. Sometimes it was one calling the other—mostly it was the sure voice of Annie Bowersox on the other end of the line, telling Micki that she was there if she needed her. But more often, it was another astronaut's wife or a relative from one side of the family or the other, from near or far, or it was a friend, close or distant, or it was a reporter, usually clueless. Across the country, there was a small army of people who, when they turned on their televisions and saw *Columbia* fall apart, immediately cast their minds toward station and the men inside of it and very soon after to their wives in Houston. Now those people almost always began their conversations with the same question: "How are you?" It didn't require nearly as much thought to ask as it took to answer.

Annie told them she was fine, that she knew in her heart that Ken would be brought home as soon and as safely as possible. She believed in the system that she had been a part of for such a long time.

Sometimes, although surprisingly rarely, Micki would allow herself to feel a little less sure about how things might work out. In those moments she would feel sorry for herself. But soon enough she would think of *Columbia*'s wives, each of those women who had lost hold of their husbands forever and not just for a little while longer, and she would breathe out those few traces of self-pity and tell herself that she was lucky or relieved or—"Fine," she would say.

After a few more minutes of gentle hand-stroking across the wires, she would hang up and wait for the phone to ring again.

Still, she was topped up with worry. About Don and when he might come home, but more about Evan and Garrett. Both of them had started showing early signs of asthma. They were constantly sick in the night, up coughing, and no one was there to spell her. She was exhausted. Her dark eyes grew circles under them to match.

Finally, after a sleepless, stress-filled month, after the phone had rung for the thousandth time—reminding her that no matter how hard she had tried to forget, all was still not right with her world—she broke.

She had been asked to give a talk to local couples who were having trouble making a baby. In the way Annie Bowersox had given comfort sprung from her hard-won experience, now Micki Pettit mined her own difficult history to make strangers feel better about their futures. But in that mining, she had to dig through six years of painful memories: the drugs, the needles, the failed in vitro attempts. Finding out she was finally pregnant, but finding out alone. Wrestling through a challenging pregnancy that she dreaded losing. Don rushing home when she gave birth to premature twins. Six weeks of an almost constant vigil in intensive care. Don having to head back to Star City only two weeks after the boys had come home. His unexpected assignment. The near agony of waiting for liftoff. Thanksgiving, another birthday, another Christmas without Don. Her boys coughing through the night. *Columbia*. And now a phone that wouldn't stop ringing, and a long line of people wanting to ask her how she was.

"Fine," she had said again and again. And it was true: somehow, she had made it through all of it, every last brutal bit of it—at least until she had finished her talk, carried the twins back to her car, strapped them in, and climbed into the front seat. Then, before she had the chance to push the key into the ignition, her tears started to flow. She hadn't seen them coming, but now she put her forehead on the wheel and allowed herself a good, hard, cleansing cry, letting out months of pain and fatigue and upset. She let everything out.

That car, in that parking lot, didn't look anything like the booth at the back of a church. And her kids—unknowing and hungry in the backseat—were her only audience. Still, in that moment, after she had wiped away the last of her tears in the rearview mirror, Micki was absolved. She had found the peace that Annie Bowersox and her three kids had, knowing that as often as it disappeared, that small, steady white light would come up once again over the trees, crossing their starlit sky on a smooth, predetermined path, the same path that would carry it over South Africa by the time they went to bed.

The crew of STS-113, including Expedition Six, board the Astro-Van on their way to the launchpad. Ken Bowersox is in the foreground, flashing the victory sign. Nikolai Budarin is behind him, smiling at the camera. And Don Pettit is the last of the astronauts, waving to reporters and friends.

After a series of delays, the space shuttle *Endeavour*, with Expedition Six stowed mid-deck, finally lifts off into Florida's night sky on November 23, 2002.

Expedition Six poses for a family portrait shortly after their arrival at the International Space Station on November 26, 2002: (from left) Nikolai Budarin, Don Pettit, and Ken Bowersox.

Don Pettit's "Saturday Morning Science": A thin film of water held in a loop of wire reacts to being shaken in weightlessness.

Don Pettit, Expedition Six's science officer, uses his trusty Makita drill to fix the troublesome Microgravity Glovebox inside Destiny, the American lab.

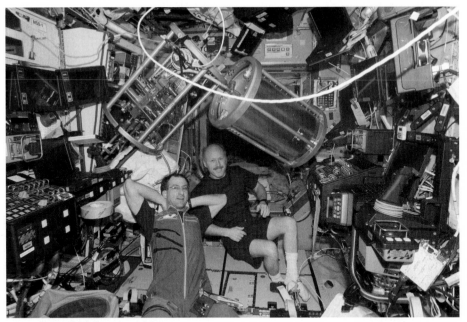

Don Pettit and Ken Bowersox, nearly lost in the clutter inside Destiny. Floating above them are the supply tank and pump for the Internal Thermal Control System.

Nikolai Budarin, Expedition Six's Russian flight engineer, is pictured inside the relatively orderly Zvezda, the heart of the International Space Station.

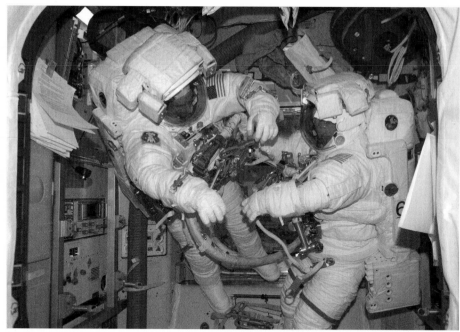

Don Pettit (left) and Ken Bowersox, trying on their spacesuits inside the Quest Airlock, in preparation for their first spacewalk, in January 2003.

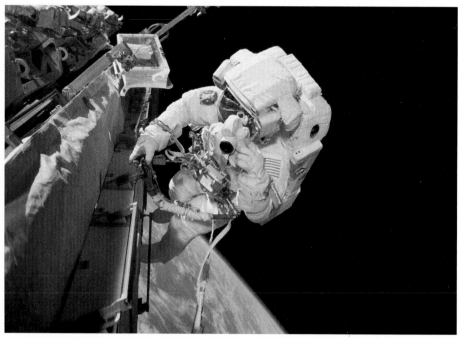

Don Pettit snaps a photograph of the P1 truss after he and Ken Bowersox put the finishing touches on its installation.

Columbia's lost crew: (left to right) Dave Brown, Rick Husband, Laurel Clark, Kalpana Chawla, Michael Anderson, Willie McCool, and Ilan Ramon.

On February 4, 2003, President George W. Bush speaks at the memorial service for *Columbia*'s crew: "We lost them so close to home." Sean O'Keefe, NASA's administrator, is seated to the president's immediate right.

In April 2003, inside Zvezda. Nikolai Budarin (left) and Ken Bowersox pull on their Russian SOKOL spacesuits, practicing for their return flight to earth on the *Soyuz TMA-1.*

A long fall down: Don Pettit slips out of the Quest Airlock at the start of his second spacewalk, in April 2003, the Earth glowing beneath his feet.

Expeditions Six and Seven, finally united, in Zvezda. Ken Bowersox and Nikolai Budarin are in the back row. Ed Lu, Don Pettit, and Yuri Malenchenko float side by side in front.

As seen through a window on the International Space Station, the *Soyuz TMA-1* capsule—carrying Ken Bowersox, Nikolai Budarin, and Don Pettit—begins its harrowing journey back to earth.

A *Soyuz* capsule attached to the International Space Station's hull: the spherical orbital module, the bell-shaped descent module, and the cylindrical propulsion module, complete with solar panels.

The full length of the International Space station; toward the bottom right corner of the photograph, a *Soyuz* capsule is attached to Destiny. The Caspian Sea provides a breathtaking backdrop.

6 THE BEST PARTS OF LONELY

Since forever, it's always been easier to be the one who's gone away. There aren't the same feelings of having been abandoned, and there aren't the constant reminders that someone is missing from life. Along with everything else that the departed leave in their wake— the unfinished book on their nightstand, the growing pile of un-opened mail on the kitchen table—the uncertainty is perhaps the worst of it. For those left behind, it makes for empty days and rest-less nights, hours occupied by asking the same question over and over again: *When will you come back home?*

After *Columbia*, Expedition Six were finally forced to share mystery's burden. They knew that they weren't coming home some-time in March, as had been planned. And they knew—despite reas-surances from the ground—that they wouldn't return to earth on *Atlantis*, the shuttle that was meant to retrieve them. Beyond those absolutes there was only a seemingly infinite list of unknowns. And as hard as Ken Bowersox, Nikolai Budarin, and Don Pettit tried to block out the questions of how they might return, when they were bundled up in their sleeping bags late at night, they couldn't keep their minds from groping in the dark for anything that looked like an answer. They never found one.

They weren't alone in their sleeplessness. At the Johnson Space Center in Houston, at the Kennedy Space Center in Florida, and at NASA's headquarters in Washington, D.C., hundreds of men and women were troubled by the fate of these three astronauts who had so suddenly lost their ride home.

First among the insomniacs was Sean O'Keefe, NASA's relatively green administrator, only the tenth chief in the agency's history. Within the Johnson Space Center's walls, he had been a controversial appointment, an astutely political, numbers-first, professional bureaucrat. After graduating from Syracuse University with a masters in public administration, O'Keefe began his civil service ascent as a presidential management intern. He later joined the staff of the United States Senate Committee on Appropriations and became staff director of the Appropriations Subcommittee on Defense. A burgeoning reputation for budget consciousness earned him the post of comptroller and chief financial officer of the Department of Defense under President George H. W. Bush in 1989. Three years later, he was future vice president Dick Cheney's secretary of the navy. When the next president Bush tapped him to head NASA, O'-Keefe had been serving as the deputy director of the Office of Management and Budget. One of his last acts before joining NASA was to reject the agency's request for $5 billion in emergency funding to balance shortfalls in financing the International Space Station. It was an alarming introduction for longtime NASA staff, who harbored no illusions about their new boss and his priorities.

But despite his bottom-line sensibilities, O'Keefe gradually won for himself a somewhat warmer reputation. He was a big man, born in down-home Louisiana of Irish descent. That alone almost guaranteed him a presence in rooms and corridors. It also guaranteed that he was demonstrative, emotional, and unexpectedly candid in his speech, despite having spent his career tightrope-walking. O'Keefe sounded more like a pilot than a pencil neck, with a deep, easy drawl that spun out *hell*s instead of *heck*s. Even on those rare occasions when his voice didn't carry out in front him, announcing his arrival, he was one of those people who's easy to spot from a distance: lumbering, always seeming vaguely uncomfortable in the suits that his station in life forced him to wear, with a head of silver hair and a thick gray mustache.

O'Keefe was hard to miss, and after he was confirmed by the Senate and sworn into his new post on December 21, 2001, he began injecting himself into every aspect of NASA's impossibly dense operation. He would never know much about rocket technology or the substance of the experiments that his charges conducted, but O'Keefe compensated for that with his time-won knowledge of people and how they operated. He took pains to remember names and faces, to make a phone call when he could have dashed off an e-mail. More than anything else, he prided himself on being visible, available. He prided himself on *being there.* And so it was that on the morning of February 1, 2003, O'Keefe had been standing in the sunshine beside that runway in Florida, laughing with the families who waited for their husbands and wives to touch down in *Columbia.*

He didn't have an inkling that something was amiss until a little after nine o'clock, when he saw Bill Readdy, a former astronaut-turned-manager and one of O'Keefe's principal advisers, walking toward him, looking ashen, and holding the dreaded contingency handbook. It was the sort of thing that was kept in a glass box with a hammer hanging next to it. O'Keefe saw the look on Readdy's face, and next he saw the handbook, and the bottom dropped out of his stomach.

"We should have heard the sonic booms by now," Readdy said after he had reached O'Keefe's side. "This can't be good."

Shortly thereafter, the families were hustled away. O'Keefe and Readdy and a few others stayed put, watching the countdown clock tick down to zero. It only confirmed what they already knew. And yet some irrational part of them had needed to watch the passing of the moment of *Columbia*'s scheduled return, uninterrupted by the sound of rubber burning on the blacktop, before they could decide what to do next. Some part of them had needed the hope-snuffing confirmation of that empty runway. Now certain that the show was over, they could begin the monstrous task of picking up the pieces and sorting through the wreckage.

Almost immediately, thoughts turned toward Expedition Six—where today would leave them, as well as tomorrow. In a break from past practice, O'Keefe had been in touch with Bowersox, Budarin, and Pettit throughout their mission. He had called them and their wives every couple of weeks, just to let them know that they hadn't been forgotten. O'Keefe had come to know the men not as friends, exactly, but better than most of the inputs in his enormous catalog of names and faces. On Christmas, he had even sneaked in a quick call to station from his kitchen.

He called them one more time, less happily, not long after worst fears were confirmed. Bowersox, Budarin, and Pettit nodded first when he offered his condolences, and next when he said that he had dispatched comfort to their wives, and last when he promised that he would bring Expedition Six home soon.

Like them, however, he just had no idea how.

. . .

Filling in the gaps in their conversation with O'Keefe, the three men knew only that they might as well settle in more deeply. If they weren't going to be back in their first homes anytime soon, they needed to make the best of their second, locking themselves inside their own private Idaho, spotless and serene.

Keeping up his crew's morale became the principal mission for Bowersox. He saw the International Space Station as a big, beautiful ship, and as its commander, he felt that his primary responsibility was to keep it afloat—not only by keeping it in orbit but by keeping it buoyant. More than anything else, he wanted Expedition Six to be remembered for its harmony, for helping to prove that it's possible for three entirely different people to live together in a tin can, under stress, and still get along like old friends.

They had been told by station veterans that their first month would be the easiest, when they would be mindful of one another's feelings and opinions, like the early days of a marriage. Then, rising to the surface like driftwood, their true selves would come out, once they were homesick enough and tired enough and bored enough for their tempers to boil over. But Bowersox, Budarin, and Pettit had found the opposite was true. With each day that dawned between them and Columbia, they became only more comfortable with one another and their shared predicament. Station grew only more into home.

They came to appreciate how their days unfolded exactly as

they wanted them to. They liked never having to alter their routine to make room for someone else in it. They were never caught in traffic or in the rain, bumped into on the sidewalk, jostled on the subway, tied to a desk for hours each day. They never caught colds. They never had to keep appointments or cut the grass. They were never rushed. They were never late.

They also came to trust one another in ways that they had never known before, the sort of unspoken trust that comes only with the knowledge that from here on in, they were on their own. Once they had found that, just about everything else fell into its one best place. Their lives were a strange, unnatural kind of perfect, almost cloud-like. Every day but one had begun with the first of sixteen dawns and the promise of peace.

In the days after disaster, the men resolved to find peace once again.

The healing really began when Expedition Six held their first audience with reporters on the ground. They were asked how they had made it along since they had learned of the loss.

A soft-spoken Ken Bowersox answered for the group: "Well, the folks on the ground have been real good about reducing our schedule, and we've had time to grieve our friends," he said. "That was very important. When you're up here this long, you can't just bottle up your emotions and focus all the time. I mean, it's important for us to acknowledge that the people on STS-107 were our friends, that we had a connection with them, and that we feel their loss. After the memorial service . . . it was very, very quiet on board the International Space Station. But now it's time to move forward, and we're doing that slowly. This press conference today is a huge step in helping us move along."

What Bowersox didn't say, but what the three men had learned, mostly on their own, was that there was some power in space that had intensified their emotions, the good and the bad equally. Whether it was the luxury of the time that they had to look inside themselves, whether it was the lumps that caught in their throats almost every time they passed by a window, whether it was a manifestation of their extraordinary loneliness, they didn't know. But they were surprised by how long it had taken them to stop their flow of tears. They were professional astronauts, and they had jobs to do, and they had trained for years for every contingency, including bad news from the ground. And yet their eyes had filled over and over again with great pools of water that wouldn't fall, and it had seemed for a long time as though nothing would break the sadness, as unshakable as the silence inside station.

Slowly, though, starting with that press conference, they began coming out of it, their collective funk lifting like a fog. No matter how forgiving Mission Control was, they were, ultimately, helped along by having so much work to do. They also took breaks to listen to music with happy memories in it, and they distracted one another with stories and questions, and they savored every consoling phone call from home. But they were helped most by the universe. As low as it had sunk them, now it began lifting them back up, on their way to heights that those of us who have spent our lives grounded can scarcely begin to imagine. Looking down on earth had never made them feel insignificant or small. Always, it had made them feel as though they were standing on the shoulders of giants, and now they started to make their way back up to their lookouts again.

. . .

It hadn't hurt that a loaded-down *Progress* soon arrived, stuffed full with manna from Heaven on earth. Expedition Six gathered around the hatch, like children surrounding the tree on Christmas morning, and together they opened it up.

Living in a place that could smell like an auto body shop, like so much metal and grease and sweat, they were nearly dumbstruck by the sweet aroma of fresh fruit. The Russian support crew had topped their shipment with oranges, apples, and lemons, and that bushel was the first beautiful goodness that Bowersox, Budarin, and Pettit laid their eyes on. They grabbed for ripe citrus and held it close to their faces, breathing it in, a wide-smiling Bowersox and

Pettit posing for photographs, having turned their oranges into clown noses.

Beneath the fruit, there was literally a ton of food, hardware, batteries, water, replacement parts, and gear for new experiments. There were also care packages from Houston and Moscow, bundles of gifts put together largely by their families. There were notes, cards, and artwork from their wives and children. There were home movies, their young boys splashing in swimming pools. There were candy bars and good, dark chocolate. For Bowersox, there was a tube of garlic paste. For Pettit, there were some books, including one filled with his favorite poems by Robert Service.

That night, tucked away in his sleeping bag, he leafed through it by a thin light. He had read each of the poems before, but now, after what he had been through and given where he was, they took on a new meaning for him. He was especially struck by one titled "The Men That Don't Fit In." One stanza starts:

If they just went straight they might go far;
They are strong and brave and true;
But they're always tired of the things that are,
And they want the strange and new.

In that, and in the rest of it, Pettit saw much of himself, all the more so having spent the day unloading *Progress*.

For Expedition Six, their first job after *Columbia* proved to be a therapeutic exercise as well as a transformative one. With every bag and bundle from their old world that they had floated from the ship and into their station, they found more room for the truth of their new lives to sink in. It was as though a raft had washed up on the shores of their desert island, and they saw everything they had won and lost in all that it had carried. And when they were finished with it, and after they had pushed it back out to sea empty—as though they had made the choice to stay on their island rather than try to pole their way home—they said goodbye to what they had been before its arrival. In some strange way, *Columbia* had started their rebirth, and now *Progress* had finished it. They were changed men.

. . .

But with morning came a starker realization. As happy as they were for their new supplies—if they were careful, they agreed, they might be able to stretch out their food and water until June—the full galley and drink bottles reminded them that they could just as easily be emptied.

From station's earliest days, the shuttle was its principal barge—ferrying not just its modules and parts, but also everything else that its crews needed to survive. Even stuffed to the hatch, *Progress* couldn't carry a quarter of the supplies that a single shuttle could haul. Bowersox, Budarin, and Pettit didn't need an abacus to do the predictive math. For as long as the shuttle fleet was grounded, each of the three men was going to need to live gently, and even then, *Progress* couldn't be relied on to keep them fed and watered indefinitely. They weren't yet on a convict's rations, but there would be no more holiday feasts, and for Pettit, there would be days without coffee. Their daily lives would need watering down, if only there was water to spare.

They also had to hope, more than ever before, that nothing went wrong with their ship. If a sizable part broke down and needed to be sent up in the next *Progress*, there would be less room devoted to their own needs. As much as they had come to love their new home, they feared that they might one day have to fight with it for the ground's attention, and in that respect, their collective fates had just been tied to their machine in more ways than one.

. . .

Overnight Expedition Six had become the second kind of science in space. The first is programmatic science—those studies that have been planned sometimes for years, experiments in fluid dynamics or crystal growth or protein production. The second, what Bowersox, Budarin, and Pettit had become, is fluke science, the science of accident.

None of them was a stranger to it. But now Pettit, in particular, distracted himself by opening random doors in the hallways of his imagination. He no longer confined his experiments to Destiny or to the racks of rudimentary physics and chemistry. He began to look at the entire station as his once-in-a-lifetime laboratory, and everything in it as an object of curiosity. Everything, he decided, had secrets to share.

Solving water became one of his principal riddles. He had stumbled upon his interest in fluid dynamics in weightlessness innocently enough after he grew tired of drinking everything through a straw. First, he began squeezing out shimmering spheres of coffee into the air, where they'd wait for him to swoop in and collect them in his mouth. Soon, though, he realized that he could play his own version of catch and release, pinching the spheres between his chopsticks and popping them down his gullet, where they would splash against the back of his throat like water balloons. For a long time after, he ate his drinks—one more thing that he did because he could do it up here, and he couldn't down there.

But, as it usually did for Pettit, the play turned into questions, which yielded to the hunt for answers. Although the three men kept up with most of the programmatic science they had been assigned— Budarin the mushroom picker especially liked tending the green pea plants—Pettit found the time for his own brand of research. He pinned up a backdrop comprised of a white towel laid against a dark blue shirt, set up his video camera, and filmed himself turning water into art.

Early on, he created thin films by slipping a wire hoop into a bag filled with water, like those plastic rings with which children blow bubbles. The result was an intricate little window only three hundred microns thick; in essence, he had made water into a two-dimensional object. He could blow on the films and shake them and bounce them and see what they did in response, but that carried his interest only so far. He began injecting food coloring into the water or mica flakes or salt crystals. They would dance in beautiful, unexpected patterns whenever he stirred them with a syringe or heated the water or shone a flashlight on it. After he grew a little braver, he punched his hot soldering iron through the windows, and the water sizzled and bubbled but never lost its shape, throwing off steam and tiny, boiling droplets in every direction. In station, he could use the same ingredients to make coffee or fireworks. Suddenly the possibilities seemed limitless.

Soon he graduated to great spheres of water the size of soccer balls, anchored by the same wire hoops. If he blew on them, gorgeous wave patterns crossed them, swinging back and forth in a seemingly endless loop. If he tossed an Alka-Seltzer tablet into them, they frothed and danced into a white globe that would eventually explode itself into a thousand droplets. (He discovered that if he began to spin it before it self-destructed, however, the bubbles were confined to the center of the sphere, forming a perfect white axis.) And if he used his syringe to make an air bubble inside the sphere and then filled that pocket with droplets of water, they created what he called his "symphony of spheres." Taken together, they looked a little like rain falling inside itself, until the larger sphere gobbled up the smaller ones, winning some mysterious battle between mass and velocity.

There was a fundamental beauty in each of these tricks. For Pettit, that was reason enough to perform them. But at night, tucked away with his computer, he would watch his day's work in slow-motion and wonder what made his inventions do the beautiful things that they did. He knew that there was science locked away inside his art, some practical application just waiting to be lifted out of the water. Somewhere in the middle of his symphonies were the answers to thunderstorms and interplanetary physics, the sorts of eurekas that keep men like him up for days at a time, frantic.

He didn't stop there—he couldn't. He made centrifuges out of old shampoo bottles, watching their contents collect around the sides, leaving a perfect, circular void in the middle. He set up a metal platform and threw bolts at it, at different velocities and rates of rotation, trying to predict how they would bounce back and almost always finding that they disobeyed him. And he filmed himself spinning just about everything he could lay his hands on—books (including *Understanding Engineering Thermo* by his old professor,

Dr. Octave Levenspiel), empty eggshells, camera lenses, water bottles—just to show that in space, everything charted its own course. Everything followed its own orbit through the universe.

. . .

For all three members of Expedition Six, it was the travels of light that they liked watching the most. On earth, light had seemed a simple mechanism: when the sun was out, it was there, and after dark, it was not. But from the vantage of space, through their breath-fogged windows, Bowersox, Budarin, and Pettit saw light do so much more than flick itself on and off. They saw it dance and swirl, change color, and turn into liquid, smoke, and sometimes a mirror. For them, light became a changeling.

They especially enjoyed its company at dusk. Pettit could break down the physics of twilight and explain why they were seeing what they were seeing, but for once, the why of it didn't matter, even to him.

When the night wrapped its way around the earth, its leading edge looked not so much like a clean, defined band, but more like rolling surf. Expedition Six could see thick, curling waves flooding out the remains of the day—mostly green, but sometimes red, and depending on the phases of the moon and the cloud patterns that they covered up, some blue and a hint of yellow might be mixed in just for gasps. It looked as though the sunset was guided by the tides rather than the other way around.

Minutes later, the view became even better. After night had made its full pass, it left airglow in its wake. Through the window, the earth's true horizon curved in the distance, smooth and jet-black; above it, the atmosphere remained lit up somehow, by a kind of thinly spread twinkle. That milky layer had something to do with atomic oxygen and its reaction to the billions of solar particles raining down on it, but the men of Expedition Six forgot all about that when they put on their headphones, waited for the strings to come in, and watched stars shine through the earth's own light.

And still there was more, especially whenever station passed over Canada. That's when they saw aurora borealis, the northern lights, rising high over the Arctic, swirling around the magnetic north pole like a multihued hurricane. It changed its shape and its color with every pass, sometimes looking as dense and foreboding as thunderheads, sometimes looking as delicate as breath that had frozen in the cold. There were nights when it looked as though it might produce its own soundtrack, when, if only they could crack open their windows, Expedition Six might hear the light howl.

During blessed orbits like that, the men forgave themselves for feeling teary and sentimental, as though they were listening to wedding speeches or a few hundred cellos rise up at once. Moments like those—when it seemed as though the earth and its wonder was there for them and for them alone—made them never want to come down.

. . .

But the moments in between had little room left for dreaming. They were filled instead with the sort of reality that feels cold and unforgiving, as grimly opportunistic as termites. It would wait to confront them until they were quiet and alone and their window was out of their reach. It knew when their minds were ripe for invasion.

It attacked most often after the ground had delivered the latest findings in the *Columbia* investigation. Each snippet of news brought back another somber recollection or planted a new seed of dread. The bad feeling reached its zenith when Bowersox and Pettit woke up to find that a short length of film had been uplinked to their computers. The footage was grainy and blurred by heat shimmer, but slowed down enough, it captured what looked like a piece of insulating foam breaking off *Columbia*'s external tank during its liftoff and striking the underside of its wing. Maybe it was nothing. Maybe it was everything.

The men of Expedition Six watched the sequence until they had committed every frame to memory. It became their version of the Zapruder film—and like those who had turned away when Kennedy's head had snapped back, Bowersox and Pettit always shuddered at the moment of impact.

Foam had often fallen from the tank (during Expedition Six's own launch, fragments had peppered *Endeavour*'s belly hard enough to crack its heat-resistant ceramic tiles), but it had never wielded the heft to damage the shuttle fatally. Now, although there were still engineers and technicians within NASA who continued to dismiss the theory, it looked more and more plausible. It was finally confirmed when a piece of foam was fired out of a gun at a rein-forced carbon-carbon panel, replicating the collision, and it made like a cannonball. Looking at that entrance wound gave every astro-naut the feeling that they had cheated death only because the aim of their own lost foam had been less true. It was as though they had each been lucky enough to duck bullets, but their friends had not.

For the men on station, having survived the trip up and having yet to make the trip down, the realization left them swallowing a hot, sick feeling. They suffered from the sweats that follow catastro-phe averted, the closest of calls. Perhaps sensing their discomfort, the ground told them not to worry, the foam problem would soon be fixed, the shuttles would return to flight, and before they knew it, a crack crew would be knocking on their door, ready to bring them home.

But Expedition Six knew that was so much wishful thinking. They knew that the shuttle was still a long way from returning to space—months, probably years. And even if the fleet did launch again, it would carry up with it a prohibitive failure rate. Despite the best efforts of its greatest defenders, its reputation was sealed. Of the five shuttles, two had been reduced to heartbreak. Expedition Six weren't in any rush to hitch a ride back to earth on what might prove to be the third.

. . .

Instead, they spent their time busily turning their desert island into more of a paradise than a prison. Together, they rededicated them-selves to making station into a sanctuary, and to figuring out how they might stay hidden away for a long time in it.

They began to experiment mostly on their own bodies, which still, despite decades of advances in weightless exercise and psychological support and their own pure will, represented the greatest single constraint to interplanetary travel. We are well on our way to building the machines that will carry us to Mars and beyond; we just have to find a way to help our fussy, fragile bodies catch up.

Muscle atrophy and bone decay remain the biggest concerns, the loss of the strength that rising out of bed every morning gives us. But long-duration flight presents more subtle dilemmas, too: the long-term effects of carrying too much blood in the head and chest; the rampant growth of kidney stones; nausea; a loss of hand-eye coordination; trouble sleeping and fatigue; and the development of a host of immune system deficiencies evident upon an astronaut's return to earth, a consequence of so long spent in a germless bubble. There are also the psychological problems that can surface over such a long time away: deep feelings of loneliness and depression (such as those John Blaha experienced on Mir), irritability and moodiness (evidenced by the Skylab mutiny), and even a brand of acute paranoia, the sort of sweaty claustrophobia that science-fiction films have made into a virtual cliché.

The mysterious case of Salyut 5 is the classic example. Having been left empty for two weeks, the capsule was occupied by a pair of highly trained cosmonauts, Boris Volynov and Vitaly Zholobov. It was expected that they would remain in orbit until they broke the Soviet space endurance record. But after just seven weeks in space, they were called back down to earth and hustled out of view. It was never made public at the time, but now it appears that the crew began to break down in the isolation. It started with minor quarrels between them and the ground, escalating into a series of complaints over several unverifiable plagues. The final straw was their repeated bawling over an inescapable, acrid odor in the crew quarters, the source of which could never be found and which could never be fixed. It was also never a problem again.

Home and away, there are countless links between physical woes and psychological troubles, but space seems to be a particularly holistic environment. There have been frequent cases of the mind following the body down a slippery slope, astronauts and cosmonauts having been driven nearly insane by complaints of heart murmurs and shortness of breath, ailments that disappear as soon as they're back on the ground.

So, like foundation inspectors trying to ferret out the rot, flight surgeons and medical technicians have used the International Space Station's crews as subjects in a grand experiment, dedicated to improving the physical and mental well-being of long-duration astronauts, in the hopes that one day the men will learn how to keep themselves as fit as their machines. They are so heavily monitored that they can feel as though they're on the slab and about to have their ribs cracked open, living subjects for an autopsy.

All three members of Expedition Six were dissected, but in some ways, Ken Bowersox was their principal cadaver. In tremendous physical shape from his years of military service—he has been a dedicated jogger since his days at Annapolis—he was an ideal candidate for learning how best to keep the human body from failing. Whatever physical functions could be tested and tracked (his heart rate, his lung capacity, his blood composition, his reflexes) were held up to the light throughout his months in space. But for most of his journey, the focus remained on the fundamentals of his architecture, the root of just about every collapse in the history of manned space flight, muscles and bones.

As recently as Mir, it was found that some cosmonauts—some of the strongest, fittest men in Russia—lost bone mineral ten times faster than do postmenopausal women. Although it's hard to imagine, bones, like coral, look rock solid but are, in fact, dynamic. They are in their own way elastic, growing and shrinking depending on the forces applied to them. Weightlessness disrupts the balance because, suddenly, an astronaut's skeleton has virtually no pressure applied against it. Like the biceps of a bodybuilder who stops lifting weights, the bones quickly begin wasting away—the great fear always having been that they will become so brittle, they will prevent an astronaut from returning to gravity's crush.

Before he was launched into space, Bowersox's bones, as well as those of the rest of the crew, were measured in every conceivable way, over several days, to determine how the loads on them changed over the course of their usual routines. Once on station, Bowersox was again monitored, with particular attention paid to his legs. Every so often, he slipped into what the flight surgeons called a "lower-extremity monitoring suit." Really, it looked like a pair of black tights with sensors at Bowersox's hips, knees, and ankles. He also put insoles into his running shoes and hooked himself up to a portable computer that recorded the data.

Not surprisingly, it was found that a day in his new, weightless life had fewer burdens than the ones he carried on earth. Even when he pushed himself as hard as he could on the treadmill or pulled a hundred squats in the Node, the force was nothing like what he'd reach during his evening jogs in Houston. It was as though he were running on air, leaving technicians on the ground with the hard knowledge that they still had work to do. If they were going to make Mars possible, they needed to invent new and better exercise machines and programs for the astronauts to follow. Otherwise, over the months and even years, the men and women who had been sent away might risk coming back in pieces.

Except that Bowersox never started fading away. Neither, to any great extent, did Don Pettit. Their bodies disobeyed not only math but also physics and logic.

Because they didn't know how long they might remain in space, each of them spent at least two hours a day exercising, sometimes at fiendish paces—but the numbers that were being fed to the ground indicated that no matter how much they exercised, it probably was never going to be enough.

Instead, every part of Bowersox, especially, remained fit, and in some cases fitter than he had been at home. Against all reason, it appeared that when he finally returned to earth, he would hit the ground running.

For all the head-scratching by the flight surgeons back in Houston, it soon became clear that there was only one possible explanation. Maybe the body didn't lead the mind after all, or at least not always. Maybe it was the mind that led the body. And because the men of Expedition Six were bent on being as happy together in space as they could imagine being on the ground, their bodies had been fortified along with their spirits. Never mind the readouts from the sensors strapped to their legs and the physics of breakdown. Maybe it was simpler than that. Just maybe, because they were having the best time of their lives, they were in the best shape of their lives, too.

. . .

In a way that they never had on earth—its landfills jammed with cheap electronics, junked cars, and the latest and greatest in entirely disposable crap—the men of Expedition Six began to feel that everything, starting with their bodies on up, could be built to last. They looked at their small, self-contained universe as a kind of refuge from our throwaway culture. In a place where even the heat of their breath was put to good use, they came not only to accept conservation as a way of life but also to take pride and pleasure in it. They became *environmentalists* in the truest sense of the word, hyperaware and vigilant in their housekeeping. Without room to sprawl, they won an intimacy with their surroundings greater than they had ever known down there, especially in a big city like Houston, especially in a big state like Texas.

One afternoon Pettit glanced at his NASA-issued wristwatch and saw, to his chagrin, that it had broken. Designed to withstand the rigors of space, it had packed it in all the same, just like everything else, shedding a button that had now become a miniature satellite. Budarin's also fell apart.

Two weeks later, Pettit found the button for his, wedged tight in a ventilator that he was cleaning. He looked at the button, and he looked at the watch, his mind having picked up on the ticking where the watch had left off.

Like Picasso going through his Blue period, Pettit had once been obsessed with clocks. He had filled days by pulling timepieces apart and putting them back together, falling very much in love with these tiny, complex engines. Although the books and manuals sent up from the ground advised that such intricate repairs were impossible to make in space, Pettit twisted the button between two of his fingers, its chrome finish shining under the light, and decided that it was in him—it was in the new and improved him—to find a way to bring life back to his watch.

He cleared himself a workspace and gathered up the tools that he thought he might need: pliers, a set of small screwdrivers, a flashlight. He had already affixed a loop of white Velcro to each of them, and now he laid out a larger sheet of Velcro on the table, a makeshift toolbox. He also stretched out some pieces of double-sided tape.

With the cameras rolling, first he undid each of the nine tiny screws that bolted down his watch's smooth metal back. When each screw floated loose, spinning through the air, he pinched it between his fingers and carefully stuck it to the tape. Then he popped off the back of his watch and pulled apart its insides, adding each successive component to his gummy collection of bits and pieces. Finally, he was able to return the lost button to its place, screwing it in tight. With it repaired, he put the rest of the watch back together, finishing the job by driving back in those nine tiny screws. It was, by any measure, a graceful and methodical operation, and he was proud of it. When he was finished, he held up his watch, as good as new, for the camera.

It wasn't just for show, however. It wasn't idle boasting. In his repaired watch Pettit saw a message, one that he had long wanted to send to the ground, to those fear-mongering technicians in Mission Control who had put more emphasis on the second word than the first. The men and women of the International Space Station, Pettit now told them in his own gentle way, were not tourists, and they were not ballast. They were mechanics and magicians and inventors, and all they had ever wanted was the chance to show it.

Now that his balky watch had given him that chance, Pettit was emboldened. Suddenly he felt as though he could fix the Gaza Strip, or at the very least a certain Microgravity Glovebox still sitting unused in Destiny. Maybe the ground had been wrong about it—and wrong about him—for all of this time, each of them needing only to be at opposite ends of a healing touch.

Pettit finally won permission to crack the box open like a safe, and almost immediately, he found a hint that would help him learn what ailed it: a green LED light on one of the power supplies was glowing more dimly than it should have been.

In the Microgravity Glovebox, there were two different power supplies bolted side by side, one putting out twelve volts, the other five. It was the twelve-volt supply that sported the too-faint light. After opening some circuit breakers, Pettit decided that, in fact, the twelve-volt supply was straight-up dead; whatever juice it seemed to be putting out was nothing more than bleed-over from the five-volt supply. The box's European technicians had thought only that the supply was diminished, not dead, and they agreed to send up a new unit.

Pettit installed the shining artificial heart and fired up the machine. It looked on the verge of working when suddenly it wheezed to a halt, as though one of its circuits had shorted out.

There were twenty connectors inside the new power supply. After some protracted consultation, the ground and Pettit agreed that he would disconnect them and begin an almost cruel, tedious process. He would connect each of them back up, but one at a time, run the Microgravity Glovebox for three or four hours, and see if he could find which one of them was faulty. With interruptions and a few much-needed breaks factored in, it took Pettit nearly three weeks to run through each connector. And in the end, none of them betrayed a short.

Now staring hard at this infernal machine, Pettit threw away his thinking cap and resolved to take a Fonzie-and-the-jukebox stab at it. He hooked the twenty connectors back up, gave the Microgravity Glovebox a gentle caress (it might have been a hard slap), crossed his fingers, and turned it on.

Lo and behold, it worked perfectly, and it worked perfectly in the days after, and it's working perfectly today—and some of our brightest minds are still shaking their big, oval heads about the entire episode. They've never been satisfied with Pettit's official diagnosis of "gremlins," but perhaps he knew better how the universe operates.

Either way, he had won for himself a brand-new toy.

And space had added to its collection of secrets.

. . .

Of the three men, Budarin grew the most in tune with the mysteries of the Milky Way. His previous long-duration missions had made him the most adaptable to the station's many quirks, and he had the most magic up his sleeve when it came down to solving the many problems of living in space. Bowersox and Pettit often found themselves watching their Russian friend go about his business, taking mental notes, learning by osmosis from his experience. (Given their checkered history, perhaps it's not surprising that Russians and Americans, even chummy ones, are still in the habit of spying on each other.)

The Americans were particularly struck by Budarin's mastery of zero gravity. He seemed downright comfortable in a permanent state of freefall, even more graceful and gentle-seeming than he was on earth. While Bowersox and Pettit still occasionally fought with weightlessness, it was clear to them that Budarin had given himself over to it. He had stopped trying to dig in his heels.

Whenever work needed to be done behind one of the International Space Station's countless panels—most of the walls are built in sections that can be lifted away to reveal machinery or stores— Bowersox and Pettit dreaded moving the running shoes, notepads, and tools that had been pinned to their fronts. They painstakingly shifted the stowaways from one panel to another, careful to strap each object down lest it take flight (and aim) at something important. After they had removed the first panel, finished their work behind it, and snapped it back into place, they returned the clutter to its former hideouts, piece by piece.

Budarin, however, took no such care. He removed the debris, but he didn't bother finding another place to stow it, nor did he strap it down. He knew that in weightlessness, each bauble—so long as he was careful not to give it any momentum—would hang in the air around him, as though waiting on an invisible shelf. By Pettit's eye, every time Budarin went to work behind a panel, it was as though he had immersed himself in a life-size game of Tetris, the puzzle pieces falling out of the ether and into his hands when he was ready for them, waiting for him to put them back in their rightful place.

Soon Bowersox and Pettit followed more and more of Budarin's unspoken lead, and not just in how he worked.

From his first days on station, the Russian had the peculiar habit of taking his allotted breakfast packet of strawberries (or the always delightful berry medley), filling it with water, and leaving it under a strap on the galley table until sometime after dinner, when he ate it like dessert. The Americans watched this routine with growing fascination—Bowersox and Pettit always ate their crunchy fruit when their menu told them to—until they finally asked Budarin why he waited for his helping to turn into a warm mush. It wasn't until Budarin offered a sample of his creation that Bowersox and Pettit understood. Somehow, the water and the waiting brought out the flavor in the berries, and their soft consistency made them all the more delicious. Budarin had learned that much on Mir.

From then on, there were always three packets of berries soaking in water on the galley table from morning until night, when Expedition Six gathered and raced through their dinners to get to their desserts. And each time they did, Bowersox and Pettit paid silent thanks to Budarin's shared wisdom, another lesson learned, another trick revealed.

. . .

Some of the tricks were less pleasant to perform than others. After brushing their teeth, for instance, the men couldn't spit out their toothpaste into a sink. Instead, they had to swallow it, gagging on the glamour of life as an astronaut.

Even something as mundane as laundry turned into an adventure in ingenuity, because for as long as men have looked to the heavens, how to do the wash has ranked alongside dark matter in the order of galactic mysteries. It was Bowersox who faced down the challenge the most, not because he was particularly dirty but because he had one favorite pair of blue shorts that he pulled on nearly every morning. (He had lifted them from the shuttle, which had much more comfortable shorts in its supply chest than station did.) On those mornings when he woke up and went to pull them on and decided that they would be able to stand up on their own (and not just because of the weightlessness), he headed to Zarya and began a long, tiresome process that he had invented from scratch, satisfying only in its results.

He squeezed condensate—a small amount of the water that had been pulled out of the air—from its container and into a large plastic bag, added a little bit of bar soap, and then pushed his shorts into it. With his hand, he mixed them together, working away until the shorts had absorbed most of the soap and water. He then took the shorts out of the bag, turned them inside out, and using a fluffy white Russian towel—the Russians had come up with a towel that was nicely absorbent and yet didn't leave too much lint behind, the sort of impossible dream that had kept Skylab's crews raging—he patted them down, looking, whenever he lost his footing, like a man trying to wrestle down two flags in the wind. After a thorough rubbing with the large towel, extracting the bulk of the water and soap from his shorts, he then dug out a smaller striped towel that was reserved for a second dry cycle and wrestled some more.

Next came the rinse cycle, which required swishing his shorts and some clean condensate in the same large plastic bag. Finally, he strapped the shorts and the towels to the bulkhead for a long session of air-drying, leaving them looking like flags once again. In about three hours, the water they held had turned back into condensate and been recycled again, perhaps this time not made into tea but into fresh oxygen or a sponge bath.

One storied evening, after Bowersox had spent a good part of his day doing laundry, he arrived at the dinner table proud of his clean pair of blue shorts. Pettit was already in the galley, and, as usual, he was already playing with his food. On this particular occasion, he had squeezed out a huge ball of orange juice and now was blowing on it, seeing what new waves he could create, watching how it reacted to life on station. Even after Bowersox had finished preparing his food and strapped himself down to eat, Pettit was still at play. He filled up his lungs and gave the bubble of juice a good, strong push—too strong, it turned out. The juice rocketed across the galley, taking dead aim at Bowersox, who until then had been happily munching away. He saw the bubble at the last second, filled with liquid menace, bearing down on him like a meteor, but he had time only to raise his hands. When its surface reached his, it exploded, grabbing hold of him wherever it could find an opening. Most of it took root in his shorts; a few wayward drops splashed into his eye. Bowersox unleashed a bloodcurdling scream, grabbing his face as though someone had thrown acid into it. Pettit, frantic, apologetic, and scooting over with a towel, asked Bowersox where it hurt the most.

"Oh my gosh, I'm so sorry, Sox. Is it your eye? Is it your eye?" Pettit said, reaching out to dab at Bowersox's face.

"No!" Bowersox cried, wincing in what was now looking a little more like mock anguish. "My shorts! My beautiful shorts!"

. . .

The story of the orange-juice bomb passed quickly through the ranks at Mission Control, where ears were always open for something to break the tedium of watching station on its endless flight around the earth.

Their ears were sometimes too open for Expedition Six's liking. During one particular Internet chat session with elementary-school students, a child asked Bowersox what it felt like to be weightless. Bowersox had thought for a moment, fumbling for a good analogy, until he settled on Peter Pan. He didn't have wings, and he didn't have to put on a jetpack, but he was still able to fly. He just had to believe, and he could jump as high as he wanted to.

The next morning, Bowersox woke up, turned on his computer, and saw that some photos had been e-mailed to him. He realized his previous day's mistake as soon as he had opened the first one: his smiling face had been Photoshopped time and again onto Peter Pan's lithe, tights-clad body. The navy man and decorated astronaut had been turned, just like that, into a prancing fairy.

But even a blushing Bowersox had to admit there was some truth in the illusion. In ways that were harder to measure and impossible to quantify, the three men thriving on station continued to make departures from the three men who had ducked through its hatch three months earlier. Even they couldn't put a finger on what it was about them that was different. They weren't the sorts of things that could be seen in a mirror or charted by a machine, and they weren't as dramatic as the shift that had taken place following the loss of *Columbia* and the arrival of *Progress*. Bowersox, Budarin, and Pettit were in the middle of a more subtle metamorphosis, a change in just a few of the million little pieces that each was made of. And yet those changes were enough to make everything else in them jangle, as though parts of them no longer fit.

For Bowersox, he felt it most acutely in his dreams. For as long as he could remember—maybe even back to the time before John Glenn opened his eyes to the universe—he had dreamed he was able to fly. He had never really thought about it, but for all his life, he knew that if he flapped his arms hard enough, he would lift up off the ground and glide over rooftops. It was in him and he went with it.

Now that had changed. Nestled against his wall of water in Unity, Bowersox still dreamed of the people and places he had always dreamed about. Even in space, he dreamed of his old self, grounded, walking along beaches or driving through the desert. But now, whenever he wanted to fly, he didn't have to flap his arms anymore. He needed only to pull off the road and onto the shoulder, climb out of his car and give a little kick, and he could fly over mountaintops and look down on skyscrapers. In his dreams, as in his new life, he really was Peter Pan. He was forever weightless.

. . .

Awake, Bowersox continued to feel just as unburdened, riding a euphoria that might have been hard to predict from the ground. Although the American space program has traditionally paid little attention to the psychological health of its astronauts in space—and most astronauts have been reluctant to discuss any problems they might have had in orbit, lest it harm their prospects for future assignments—there is strong evidence that spending a long time in space can make people crackers. Weeks and months of interrupted sleep, sensory deprivation, isolation, confinement, latent danger, poor hygiene, lousy food, chronic noise and vibration, and close, permanent contact with fellow crew members . . . Not surprisingly, that wretched mix has proved fertile ground for a host of disorders to take root. While the vast majority of astronauts and cosmonauts repel complete psychological breakdown, many have suffered from fatigue, nervousness, weakness, anger, and memory and motor hiccups. In addition to the innate power of space to push emotions toward the margins of acceptability, it also tends to bring out the worst in its inhabitants. More often than not, space will expose cracks that are invisible on the ground. It can see through masks and bravado.

But in his isolation, Bowersox—as well as Budarin and Pettit—found tranquillity. His loneliness became his salvation. He had spent his entire life busy, going somewhere, working toward something, jogging or hustling or flying faster than the speed of sound. And through all of it, he had been beholden to others and to outside forces, to teachers, superior officers, ocean currents, marching orders, his wife and his children, aerodynamics, the gas left in the tank. He was a man who had done nearly everything he had wanted for himself, and in that, he was better off than billions of the rest of us, but every step of the way, there had always been another test to pass, another standard to meet. There had always been another obligation. It was as though a life even as extraordinary as his had seen a governor strapped to it, tying him into the machinery of life on earth: complex, unyielding, and not without its rhythm; yet keeping that beat always seems to come at the cost of some measure of self. To reach the next in his long string of goals, Bowersox had sometimes felt as though he had become less of his own man.

Now, despite being confined to a single light in the sky, he was as independent as a drifter, an aimless kind of free. He continued to bear responsibilities—to Expedition Six, to the International Space Station, to NASA, and to his family—but now they were remote, more abstract. In some ways, he felt like a married bachelor: he had the comfort of union and the knowledge that his wife was waiting for him but none of the compromise that comes with the day-to-day reality of a shared life. He knew freedom without the longing for anchor, routine without the misery of drudgery.

Every time he talked to someone on the ground or answered e-mails from students or conducted a press conference from space, Bowersox was inevitably asked how he coped with the disconnect: Did he miss earth? And no matter the noun at the heart of the question—no matter the place, people, or thing dangled in front of him—always, he answered with some version of no. He didn't allow himself to submit to cravings. He didn't dream of apple pie and ice cream, the sorts of desires that have driven snowbound polar explorers blind with yearning. And in the same way, for the first time in his life, he resisted submitting himself to the whims and demands of other people. He never wanted to be put in better touch with the realities of his former existence. If anything, he felt as though he and the other members of Expedition Six remained *too* linked with it, that there were too many ways for them to talk to earth and for earth to talk to them. He sometimes wished that he could turn off the radio and shut down the Internet phone and erase every e-mail, because part of him wanted to be even more alone than he was. (He and Pettit certainly stopped paying attention to the *Columbia* investigation, because nothing good could come out of it for them.) That same part of him wanted to make a sharper divide between the best parts of lonely and the worst parts of company. For as long as he stayed on station, he could be his true, original self. In space, Ken Bowersox was entirely his own man, and he liked the feeling very much.

. . .

So, too, did Pettit and Budarin. For them, the idea of going home had made its own slow turn, gone from feeling like the end of time served to more like the end of summer camp or the last days of vacation. Even with the stress of their now open-ended mission and their dwindling coffee supply, they still felt as though they had remained a crew in the best sense. Their feeling of family deepened, became more defined. Bowersox was the firstborn brother. He was reason and responsibility. Pettit was the wide-eyed kid who loved eating his drinks with chopsticks. Budarin was the weird uncle from Russia.

And yet, somehow, it worked.

Inevitably, though, the time came for them to have it out. They never came close to dropping gloves, but try as Bowersox might to hold everything together, it was impossible for the three of them to make it through all of these weeks and months without disagreement.

The trouble started when Bowersox and Pettit began imagining the conversations that were taking place back in Houston, probably behind closed doors and probably in whispers. They knew that as much as things hadn't looked to have changed on station, there were new plans being made, new futures being drawn up. Things couldn't stay the way they were forever.

During their shared daydreaming, either Bowersox or Pettit—neither can remember which of them—raised the specter that one of them would be asked to stay in space while the other would be replaced by an incoming cosmonaut flown up on a Russian rocket. That rocket, in turn, would deliver the replaced man home. They took to calling this scenario the Avdeyev Option, after Sergei Avdeyev, a Russian cosmonaut who, like Sergei Krikalev, had also endured an unexpectedly long mission: he survived 379 consecutive days aboard clammy Mir. Though the single switch-out was unlikely, Bowersox and Pettit made the silent determination that it was not an impossibility. In fact, the more they thought about it, the more they thought it was about to happen. And the more real the possibility seemed, the more they began to talk about—and, eventually, come close to arguing about—which one of them would be picked to go home.

The problem wasn't that the earth was calling out loudly to

them, or that they had grown tired of sleeping as though from a hook, or that they were hungry for something that wasn't born of a lab, or that they feared what would remain of their bodies and minds after so much time in space.

The problem was, both of them wanted to stay.

. . .

As the days passed, however, Bowersox, Pettit, and Budarin were sometimes alarmed at how much they felt themselves continue to evolve. Change started to feel like decay. Not in any of the expected ways—Expedition Six were on their way to proving that, physically at least, men can last long enough to make it to Mars—but not confined to their dreams alone, either. They could feel, in their souls, the wearing away of the calluses that life on earth had given them. It was as though their skin had been stripped off and replaced with a fresh pink layer.

One night, Expedition Six finally decided to watch a movie. Even though there is a healthy selection of DVDs on station, smuggled over time, the men had resisted the urge to fire one up. They had always told themselves that something better was being projected on the big screen outside. But even sunrises and sunsets can get old after a few thousand ups and downs and, frankly, their to-do list imaginations had been tapped out. So movie night it was. They got themselves some snacks and bags of juice, gathered around an IBM ThinkPad mounted on a rack, and cued up a movie called *Tank Girl*. It bombed at the box office, but in the meantime, it's become a cult hit among women astronauts—so much so that Bowersox and Pettit had been told that if they did nothing else while in space, they had to watch this comic book come to life.

It might as well have been playing in fast-forward. It was jagged and bright and loud, an assault on their eyes and ears. But most of all, it was violent. In the first few minutes, a man was forced to take off his boots and walk across broken glass, crunching underfoot with each agonizing step; he was then stabbed in the back with a machine that pumped out his blood and turned it into water. Dozens more men were shot, blown up by grenades, electrocuted, paralyzed, tortured, run over by trucks, pierced with arrows, and torn apart by a mutant race of kangaroos called Rippers. Even a friendly, lumbering ox took a bullet in the head. Combined, it was enough to make their fresh pink skin crawl.

And if that wasn't enough overstimulation, there was a long, drawn-out scene in a futuristic strip joint called Liquid Silver. Beautiful women in platinum-blond wigs and silver G-strings danced on the stage; Tank Girl and her sidekick, Jet Girl, made likewise, dressing up in outfits as revealing as bikinis. So many girls—luminous girls with lips and breath and falling hair. And then our heroes lifted their guns back to their shoulders and continued blasting away.

Bowersox, Pettit, and Budarin looked down at their hands, and they were shaking. Their mouths had gone dry. Their hearts galloped. Every biological stress indicator had kicked into overdrive. None of them made it to the end of the movie for fear of system failure. Together, they agreed to turn it off, to talk to one another in whispers, and to take a little longer than usual to come down before going to bed. But even after they'd tried to unwind and pulled themselves into their sleeping bags, they still trembled, like wide-eyed kids who've been told ghost stories around a campfire before lights-out.

Come morning, they had each drawn the same conclusion: despite their gut wishes, maybe they had been gone for long enough. Maybe they needed to start thinking about going home. Maybe they needed to answer the questions of when and where and how. Maybe it was time.

Because the earth had been spinning on its axis, and they had been spinning on theirs, but now they knew that they'd been traveling in opposite directions for all of this time, and they felt as though they had never been so far away.

Down there it's a relic, gone to tumble, waiting to be felled like a tree. On the western edge of the great dry lakebed at Edwards Air Force Base, a square, timber-framed platform rises high above the cracked ground. Up a creaky flight of stairs and over a railing that wouldn't withstand more than a good push, there's still a commanding view of the desert flats, with runways marked out in oil. Beyond the slicks, mountains rise in the distance set against a pale sky.

Underfoot, scraps of outdoor carpet remain pinned down by rusted staples, and a phone line, long since dead, hangs loose nearby. Once there was a red phone plugged into one end of it, in anticipation of President Ronald Reagan's probable visit, just in case he ran out of jelly beans or decided that he needed to drop a few bombs. That was only twenty-five years ago, when this platform stood at the center of the universe, and STS-001, *Columbia*'s first flight, prepared to touch down in front of it. But for a visitor standing there today in the teeth of the wind, it's hard to imagine this place was ever anything more than the easy metaphor it's become.

Until *Columbia*'s liftoff—until the morning of April 12, 1981—space had been the exclusive property of the Soviets for six long years. Closer to home, America was trapped in an even longer losing streak. Vietnam was still a too-fresh nightmare; rescue helicopters had buried themselves into the Iranian desert; Three Mile Island was a horror story nearly come true. Even iron-hard Detroit had been forced into retreat, pinned down by an unstoppable influx of cheap, reliable cars with strange names from Japan. For the first time since the Great Depression, "Buy American" had more pleading than pride in it.

Until that lit-up morning, the space program had only added to the feeling that another one of history's great empires had run its course, the last days of the latest Rome. The new shuttle was two years late, bogged down by technological failure and a flawed design and the malaise that had gripped the rest of the nation. But under cloudless blue skies, in front of an audience of thousands in Florida and millions more in classrooms and taverns and basement dens, John Young and Robert Crippen helped pull off the miracle comeback. They successfully guided *Columbia* into the first of a long string of orbits, and it felt, in that instant, as if everything might be put in its proper order once again. All that remained was the long wait for the shuttle's arrival in California. Young and Crippen had each taken out life insurance policies worth $800,000. No one was ready to celebrate until those papers had been made worthless.

A little more than fifty-four hours of national breath-holding later, that wood platform was filled, as was the wide viewing area marked out more than three miles from the landing site, as were the mountains. (The president was not among those in attendance after all; having taken John Hinckley's bullet to the chest only two weeks earlier, he was forced to watch from the Lincoln Bedroom.) Ignoring warnings of rattlesnakes and traffic snarls, more than 200,000 bird-watchers congregated. They stuck American flags into the dry ground, pulled on their EAT YOUR HEARTS OUT, RUSSIANS! T-shirts, and turned their eyes to the sky.

At ten o'clock in the morning, they heard a pair of sonic booms, after the shuttle had bolted across the Pacific coast, the San Joaquin valley, and the Tehachapi Mountains. (At Mission Control in Houston, flight director Don Puddy announced, "Room, get ready for exhilaration.") The crowds and television cameras first caught sight of the light that banked off the shuttle's cockpit windows, and then they saw its black underbelly dropping out of the blue. Across the country, factory workers shut down the assembly lines and raced into lunchrooms. Office meetings came to a halt. A fitter in a Manhattan men's shop dashed off, leaving a customer pinned up in an unfinished suit. All of them gathered around televisions and radios to watch and hear *Columbia* soar over the dry lakebed, loop back, and touch down in a cloud of dust and the wash of a mirage. Finally, again, Americans had been part of something perfect.

Just six hectic hours later, Young and Crippen, still zipped up in their blue flight suits, stepped down from a plane and into a crowd 1,500-strong at Ellington Air Force Base in Texas. They were greeted as heroes.

Young, a veteran of four previous trips into space and a walk on the moon, was enthusiastic about his latest ride. "The spaceship *Columbia* is phenomenal," he said. "It is an incredibly amazing piece of machinery. Anytime you can take something that big and put it into space and bring it back and land it, you have accomplished something just short of a miracle."

Crippen, a first-time flier, had trouble finding the words to describe the experience. "For me," he said finally, "it was the darnedest time I've ever had in my entire life."

The international response was no less effusive. Congratulations came in from the Canadian parliament, Italian president Sandro Pertini, and the Polish soccer nut who had just become Pope John Paul II. The launch and the landing had also found places on a thousand front pages and led off newscasts in a hundred languages. *The Guardian* in London wrote: "The shuttle is *Star Trek*, *Star Wars* and *The Empire Strikes Back* in life. It is beautiful, futuristic and patriotic in an era when Americans have found little to cheer about." The French business daily *Les Echos* went one step further: "The unbelievable has become reality . . . Man will go into space now as easily as he crosses the English Channel."

But it was the laid-up president who had already best summed up the measure of that almost impossibly wonderful trip.

"Through you," Reagan told the astronauts before their launch, "we feel as giants once again."

. . .

The feeling lasted nearly five years, until *Challenger* turned into pink vapor one chilly day in January 1986. Long before, the networks had each set up permanent offices in Houston, trailers anchored in the shadows of the Johnson Space Center's big-box architecture. As in the days of *Mercury*, *Gemini*, and *Apollo*, each of the shuttle's flights had been a headline event. But after the loss of seven astronauts had been analyzed and replayed until it had been turned into a spot memory, one of those rare moments in history that can be instantly recalled by tens of millions, space had lost its luster once again. Sometime during the more than two years it took for another shuttle to slip the surly bonds of earth, Americans found different things to hold close to their hearts. Like the platform at Edwards Air Force Base, the network trailers first were shuttered, and then left to rot, and, like dreams, they were finally forgotten. Astronauts who had come of age under flashbulbs suddenly found themselves working in two vacuums, home and away.

(The greatest all-time episode of *The Simpsons* centered around NASA's attempts to lure back its former audiences by sending Homer J. Simpson, American Everyman, into space. The agency was spurred to take that desperate step when its most recent launch was outdrawn by *A Connie Chung Christmas*.)

Even without *Challenger* and the wrung-out quiet that followed it, the shuttle program was probably on the verge of being shoved out of the spotlight. The national attention span is only so long, and the shuttle's seemingly endless loop of launches and landings had started to blur into a spectacular routine. The missions in between, too, weren't the sort of finite adventures—like walking on the moon—that grabbed people by the shoulders and dropped jaws. To ordinary Americans, it looked more and more like their astronauts had become glorified truck drivers, following the same routes, hauling the same boxed-up cargo, running the same (seemingly unimportant) errands. And the truth is, as incredible as shuttle flights remain even today—when you stop to think about the physics and chemistry and psychology behind them—it's hard for millions of grounded mortals to get excited about the same fresh-faced few doing backflips, chasing down wonky satellites, and taking long-range photographs of oil spills. It can feel as though it's not worth the effort to strain our necks anymore.

Part of that feeling has been by design. Since *Challenger*, NASA's brass has viewed yawns and empty press conferences as good things, today's versions of standing ovations and tickertape parades. No news is good news, everything having gone off according to boring plan. But beyond their desire for peace and quiet, they also know that if new ground is ever going to be broken, the old ground must become as well worn as donkey tracks. Moon colonies, space tourism, manned missions to Mars—none of them is possible without first making near space seem less like an ocean and more like a wading pool. That's where the International Space Station comes in. And that's why NASA desperately needs its rockets to seem ordinary, even though they never really have been, for the next horizon to appear within reach of them.

The bummer, however, is that in making itself look like another humdrum government bureaucracy, NASA has begun to act like one. It is impenetrable and slow to respond, tongue-tied and nearly impossible to get to know. For the few brave reporters who have chosen space as their beat, it can sometimes feel as though they would enjoy better access if they covered spies for a living. (How telling is it that a grand total of three reporters followed *Columbia*'s last mission from beginning to end?) Only at NASA could it seem like a good idea for members of the press office to have the acronym IMPASS stamped on their business cards, and for the press office to live up to the label.

All of which helps explain that if astronauts land on today's front pages, it's probably because they've been blown to bits.

. . .

The day after Expedition Six rocketed toward station, the launching of seven men into space failed to earn even a brief in the *New York Times*, the nation's newspaper of record. Instead, that morning's bleak front page was dominated by President Bush's stumping for war against Iraq, including speeches at rallies in Vilnius, Lithuania, and Bucharest, Romania. "The people of Romania know that dictators must never be appeased or ignored," the president said. "They must be opposed." Below the fold, there was news of a New York City transit worker who was killed when he was struck by a train, and Hollywood was keeping its fingers crossed, hoping for a lucrative Christmas season.

There remained an almost institutional silence on space, in fact, until *Columbia* fell out of it. SHUTTLE BREAKS UP, 7 DEAD was the lead story's hard-nut headline on February 2, 2003. There were eleven pages of comprehensive coverage inside, including an editorial that sang the following lament: "Once again we were jolted out of a sustained period of success in exploring the world outside our planet—a run of good work and good luck that ran so long we had the luxury of taking it all for granted. Most Americans were probably cheerfully unaware, over the past 16 days, that seven men and women were circling the planet."

Now space was back in the news, as, finally, was Expedition Six, although they merited mentioning for only two days. The morning after the disaster, at the bottom of page 25, there was a low-key story that foretold of nothing of consequence. DELAYS EXPECTED IN EXPANDING ORBITING LABORATORY, AND PERHAPS IN RETRIEVING 3 ASTRONAUTS. A NASA spokesman named Pat Ryan snuffed out what might have turned into a good old-fashioned serial drama with his blunt assessment of the crew's fate. "The reality for all the astronauts is that when you launch, the mission is over when somebody comes to get you, and it may not always be when we planned," he said. Russia's *Progress* was already on its way to the International Space Station, he hastened to add, delivering enough food and water (and garlic paste and poetry) to keep Bowersox, Pettit, and Budarin going at least through June.

Expedition Six earned a second and final mention the following day, in the middle of another twelve pages of coverage: NASA KEEPS THREE ASTRONAUTS ABOARD SPACE STATION INFORMED OF EVENTS ON GROUND. There was a stock photograph of the crew, but no words from them. Instead, Bob Cabana, now NASA's director of flight crew operations in Houston, passed along their thoughts. "They're grieving up there also, and they feel a little isolated. They want to get through the process, and it's harder for them being detached from it in space." That hardly seemed to provide a finishing note on their story, but for the time being, it did.

Stories on the investigation into the disaster continued for a little while longer, but by February 17, *Columbia* had lost its last grip on the front page. *Challenger* had occupied the nation's consciousness for two years. Just two weeks after *Columbia* had come apart, the last few of its pieces were swept away and promptly forgotten. Perhaps because it was the second shuttle disaster and not the first, or perhaps because the pictures of catastrophe weren't as dramatic this time around, or perhaps because the nation was about to go to war, the story was considered complete. All that remained was the occasional poke at the embers of a dying fire, left virtually unattended in favor of another, now just starting to throw off sparks.

. . .

Bowersox, Budarin, and Pettit were left hanging in limbo, and limbo is death for a story. The opening scenes of their adventure had the makings for something dramatic, had the potential for emotional fireworks and a heart-tugging score, but by spring, it had been drawn out for too long to sustain its audience. It had lost its momentum. There was no conflict or progress or tidy resolution in sight. It was all verse and no chorus.

John Glenn. *Apollo 11*. *Columbia*'s first flight and *Challenger*'s last. As always, but perhaps more than ever, we are obsessed with beginnings and endings. Looking at that failing platform in the desert or at those empty trailers in Houston, we've learned that the middle, we can do without. And Expedition Six—adrift, trapped in their own middle, lost behind thousands of other headlines and movie openings and battles and football games—learned that an astronaut's universe has never been so small.

8 A THIN THREAD

And if you gaze for long into an abyss,
the abyss gazes also into you.
—FRIEDRICH NIETZSCHE

By March, approaching the start of month five in orbit, Don Pettit began filling one of those rare, beautiful times in his life—when he had nowhere else to be—by spending more time staring through his window.

Because of what he knew about the planet's architecture and construction, he could see more of it than most of us might. He could see all of the geological features that mapmakers struggle to express: braided channels, alluvial fans, glaciers, crater lakes, fissures, slumps, and volcanic plumes.

He could watch clouds of every shape and combination. He took note that clouds over cold water looked different from clouds over warm water. He could see firsthand that storms really do spin in different directions depending on their hemisphere. He delighted in lightning flashes, bands of dark clouds illuminated by fingers of spark, reaching out to each other across the miles.

He could see brown rivers spilling into the blue ocean. (They always made him think of hot chocolate.) He could watch long, solitary waves rise up in the middle of a relative nowhere, deep in the South Atlantic or far off the Alaskan coast, giant walls of water that were built up until they broke over themselves, having come and gone, gorgeous, and having been invisible to everybody but him.

He watched light scattering and corona haloes. He saw how

well jungles absorb the sun's rays, dark even in daytime, and how farmland looked bright, as reflective as a mirror.

He saw meteors falling below him, a topsy-turvy sight that he never managed to adjust to. Each time his intuition complained bitterly.

He looked at the stars, suddenly seeming close enough to touch.

But most of all, he liked looking at earth in the wee hours, after dusk's waves had washed over it. Unlike on the ground, where darkness serves only to obscure, night makes the human landscape clearer from space. In daylight, even the biggest cities look like gray, indistinct smudges, like the fingerprints on the glass in front of him. At night, however, Pettit could spend hours hunting for his life's landmarks in the power grids and black, bottomless river bends until he finally became obsessed with documenting his view. He aimed his over-the-counter Nikon down at us and began taking pictures. He built a collection of more than 25,000 in all.

At first, because the speed at which he was traveling was so much faster than the snap of his camera's shutter, Pettit's pictures turned night cities into streaks of white light, like the headlights in a time-lapse photo of a busy street. His pictures became crisper when he learned to hold open the shutter and shift his shoulders in the opposite direction of his orbit, but even then his best efforts turned out blurry. He knew that he was looking at New York City, but he couldn't quite make out the shadow of a lightless Central Park or the single bulb in the harbor that he knew was the Statue of Liberty.

Not good enough. Pettit being Pettit, he put the finishing touches on a gyroscope that he'd made out of three portable compact disc players (he used it to hold his flashlight for him), and next he built a makeshift, rotating tripod for his camera. He culled an old IMAX mount, a spare bolt, and a cordless Makita drill from the clutter, and he found eureka, too: gently squeezing the drill's trigger lent his camera the perfect rotation to take pictures sharp enough to make the miles meaningless all over again.

Looking at the electric webs of Montreal and Portland, Oregon, and Washington, D.C., he could pick out the airports he'd flown into and the street corners he knew and the hotels he'd stayed in, and he could remember if the showers were hot or if he'd enjoyed a good meal there. He could make out football stadiums, docklands, and interstates, and he could make out the rest of those big, little places that we run our lives through. In the end, if he closed his eyes, he could even see his driveway, and he could feel himself easing into it, throwing his junky pickup into park and walking up to the front door, his shoes scuffling on the asphalt, his hand guided to the door by the warm light spilling out from the windows.

In those dreaming times, it felt as though all that separated home from away was a very thin thread, so thin that Pettit could snap it just by breaking open the hatch. A loosened bolt, a turned latch, a cracked seal, and he could find himself wherever he wished to.

· · ·

He had opened the hatch once before. Back in January, he and Bowersox had performed an extravehicular activity, or EVA, which is a fancy way of saying that they had put on their boots and headed outside.

Originally, Bowersox and Budarin—the veteran of eight EVAs while he was on board Mir—had been given the assignment, with Pettit expected to stay inside, manning the station's robotic arm. But in the lead-up toward exit, routine tests found that the forty-nine-year-old Budarin's heart was not entirely up to snuff. While he pedaled away on the station's exercise bike, the machines whispered into the ears of NASA's flight surgeons, snitching on what they later called "cardiac peculiarities." The Russians explained that Budarin's heart had always had a bit of a flicker, and that the flicker had always been dismissed as benign and had never caused a problem, including during his more than forty-six hours spent at the end of Mir's tether. But because this particular space walk was an American exercise—being conducted in American spacesuits and using the American hatch—the final decision rested with NASA. Despite Pettit's lack of experience and his training only once in the pool with Bowersox (although that had gone well), he was ticketed for outbound passage. The Russians staged a minor protest, but Houston remained unmoved, and now Pettit found himself twice drafted from the second team to the first.

This time around, Bowersox didn't much care about the switch. It might not even have registered through the thrill of his anticipation. He had never been outside either, able only to watch others take the dips of their lives, and for all of this time, the dark had sung out to him.

Like the call of the waters that saw heartsick sailors pitch themselves off the backs of ships to be swallowed by Mother Ocean, deepest, blackest, emptiest space has always drawn astronauts. The same desire that makes people step off cliffs with only a parachute strapped to their backs, or sink into underwater caves or under blue ice with only a coil of rope to guide them back to the surface, also fills the men and women who have touched space with their gloved hands. The adrenaline rush that comes with suiting up, opening a hatch, floating outside, and relying on a cloth tether or a length of steel cable to keep from drifting into oblivion is just about strong enough to overwhelm spacewalkers. So, too, are the more beautiful rewards of such a short, long journey—unparalleled views and a feeling of nearly perfect isolation, as though they have been placed on pedestals that no one else on earth can dream to reach. It's no wonder some have nearly stayed forever on their perch.

Most famously, before their record stay on Salyut 6, rookie cosmonaut Yuri Romanenko came within a whisper of meeting his end during Georgi Grechko's experimental space walk on December 20, 1977. While Grechko bounced around outside, Romanenko was to stay inside the airlock and monitor incoming medical data. But his curiosity won out, and he decided to steal a peek. Romanenko poked out his head, became consumed by the view, and was lulled into drift, his safety line floating slack behind him, unattached. This, he realized too late. Fortunately, Grechko saw his crewmate's desperate thrashing and leaned over just in time to grab the end of his line and reel him in.

At the last possible moment, Romanenko had been rescued from himself and the almost inexplicable pull that space holds over us. It's an attraction so powerful that America's first spacewalker, Ed White, had to be dragged back inside *Gemini 4*, finding himself the object of frantic orders first from Houston and then from his powerless commander, James McDivitt. The star-crossed White, who would later die in the *Apollo 1* fire, finally relented and returned to the capsule, but not before announcing to the world: "It's the saddest moment of my life."

Now, along with Pettit, it was Bowersox's turn for happiness. They had been carried through the monotony of their preparations by a giddiness usually reserved for children. They had topped up their batteries and made certain that their nitrogen-thrust backpacks would fire if they needed to move in a hurry, their one shot at returning to station if their tether snapped; they triple-checked every rubber seal that separated them from the front pages and changed out their carbon filters; they layered their gold-plated polycarbonate visors with antifog solution, but not so much that it might make their eyes raw. (During an earlier assembly mission, Chris Hadfield was left with tears in his eyes after they were stung by excess juice. It was a lesson that every other astronaut had learned along with him.)

Bowersox and Pettit were just as diligent in preparing their bodies for the trip. As it is for deep-sea divers, nitrogen had become their enemy as well as their friend. While it would power those emergency backpacks of theirs—called SAFER, appropriately enough—their body's own natural stores of it threatened to turn into bubbles when they opened the door. Those bubbles would course through their bloodstreams until they became snagged in their joints and bronchial tubes and frontal lobes, leaving them incapacitated with a good old-fashioned case of the bends. That potential for crippling agony was perhaps the only link between East River tunnel workers and astronauts, between those who scrape out their livings under the earth's surface and those who fly above it.

To reduce the chance of locking up their knees and elbows and turning their lungs into a hacking pink soup, Bowersox and Pettit each took turns riding the station's exercise bike for ten minutes, while sucking back as much pure oxygen—via the masks that they had donned—as they could take in. Then they beat a hasty retreat to the airlock and spent forty more minutes behind those masks, breathing deeply. After they had completed the prescribed course of inhalation, Bowersox and Pettit slowly reduced the pressure in the airlock over the next thirty minutes. They brought it down to 10.2 pounds per square inch—the same atmosphere that they would experience on top of Hawaii's Mauna Kea, about 9,200 feet above sea level. Only then could they begin wrestling into their diapers, water-cooled long underwear, and 300-pound spacesuits.

Pettit inserted metal sizing rings into the legs to stretch his suit out a little, while Bowersox made sure the membranes were wrapped tightly around his headset, which were prone to shorting out if sweat leaked into them. Each man also installed a drink bag against his chest, careful to burp out whatever air was trapped in the water. (Otherwise, their faces might have been showered with drink, with no way to wipe them clean.) They checked and rechecked the flood- and spotlights that were attached to the sides of their domes before they slipped them on and locked them at their necks. (They checked that seal more than once.) At last, they pulled on their white gloves, which made a comforting click when they had been attached properly at the wrist.

The long process made Pettit think of knights heading into battle, wrapping themselves up in chain mail and sharpening their swords. But these modern suits were made of no ordinary armor. Even something as innocuous-seeming as those white gloves were the products of years of advanced materials engineering and dream work.

On the ground, in fact, there was an entire team dedicated to making gloves that would offer sufficient protection from intense heat, cold, and cuts, while also giving the astronauts enough flexibility to complete the delicate tasks that they had been assigned. Beneath a slick outer layer of unscuffable Teflon, there were five layers of aluminum-coated Mylar, as well as a comfort layer that Bowersox and Pettit cozied their hands into. If their digits got cold—shivering was a more likely problem than a good case of the sweats—there were even individual heaters tucked into the end of each fingertip. Those heaters could be fired up at will, leaving Bowersox and Pettit with warm and toasty hands when, without the luxury of those gloves, their mitts might have shattered like hoarfrosted glass.

It was that sort of attention to detail that filled nearly six hours of their busy morning. Finally, they made certain that their safety line, a shared cloth tether, was strung between them, and that Pettit, who would follow Bowersox outside, was tied to the inside of the airlock. Once they were certain of the integrity of both lines, they nodded at each other, watched their pressure gauge needle drop to zero, and moved to push open the hatch. "I hope I don't wake up and find out this is a dream," Pettit said.

And the goddamned bastard thing was stuck.

So close to what felt like destiny, Bowersox began banging on the beast, pushing it, pulling it, cajoling it, and finally swearing at it, none of which had any effect. Pettit, worried that Bowersox was going to turn their home into a vacuum, asked whether he might take a swing at it. The hatch resisted him the way the driver-side door on that junky pickup of his had always dug in, and he remembered, and he found his touch, and the hatch opened to a wash of the sun's bright light. It had been snagged on a loop of white fabric that had come out of place just outside, the sort of thing that could be pulled away like lint on earth but in space can turn men into satellites. Shaking his head, Bowersox resolved to cut that loop and every other one like it before he came back inside.

Outside the hatch was spooled a fifty-five-foot length of thin steel cable. Bowersox unhooked himself from Pettit and onto it. Next, he took a breath. Some astronauts throw themselves out of the hole like bomber pilots, but Bowersox had already decided on making a more graceful exit. Using a handrail for leverage, he tossed his legs out into the emptiness, allowed their momentum to pull the rest of his body outside, and finally looked down between his feet at the earth. For the second time in his life—for the first time since he had last landed on an aircraft carrier—a short length of steel cable was all that kept him from a good push into a black eternity. It was all that tied him to one of his worlds and, in turn, to the other.

Although it was unlike him to do so, Bowersox allowed himself to stall on that. It was so special a moment that he gave himself permission to turn off the automatic pilot, and he took the time to take it all in:

There's my feet. There's the earth. There's my feet, and there's the earth, and there's a long way in between.

. . .

Now, should something go wrong—a snapped tether, a hand or a foot restraint breaking free of the hull, the hatch door locking shut—there were only so many outcomes. Now, in all of that wide-open space, your range of possibility was terrifyingly narrow.

It would begin, like all knowing deaths, with panic. Probably not a screaming, thrashing panic, because your years of training wouldn't let you accelerate the process like that—and because you wouldn't want the voices on the radio to sense the tremors in yours—but there would be panic nonetheless. Your heart rate would rise. Your breathing, as much as you tried to keep it slow and even, would pick up, become shallower. Despite the cold water still running through your long underwear, sweat would start coming out of your forehead, but without gravity it wouldn't fall. If any drops were somehow shaken loose, they would float around inside your helmet, like the flakes inside a snow globe, until they had gathered enough steam to splash into your visor or bounce back into your face. That's when you would taste the salt, when you would lick your lips and begin whispering to yourself, looking for angles, for oversights, hanging on to the last living moments of your reason, trying to find a way home.

Depending on when you were cut loose, you might spend as long as seven hours staring out into your own private abyss: forever, but not long enough for any of the astronauts inside to suit up, and even if they could get themselves out in time, they would have no way to gather you in. If you were set adrift by force, bit by bit the station would become another star in the night sky. More likely, you'd be stranded maybe a hundred feet outside the hatch, just out of reach, the nitrogen in your backpack lost in a misfire, and you would hang there, locked into a new orbit running alongside your old one, without ever having the chance for them to meet. All of this you would come to understand. The panic would yield to resignation, the resignation to grief. You might pass along some last wishes. You might ask the ground to play your favorite songs. You might just turn the radio off. When the air began to sit too heavily to ignore the weight of it any longer, the most strong-willed astronauts might open up the two emergency oxygen canisters strapped to the bottoms of their packs. That would give them another hour to say goodbye. But like most of them, you wouldn't. You would just tell everybody you loved them and choke back the tears and let inevitability wash over you.

Suddenly the life would really start draining out of your blood. You would start to tire, as if you were on the final leg of a long flight. It would be hard for you to know it, but your lips would start turning blue. Your fingernails, too. Then your vision would start to fail. It might become fuzzy, or you might see two suns in the sunset, or you might find yourself unable to focus on anything outside of a single point in the distance, near or far: one of those drops of sweat set loose inside your helmet, the light reflected off a solar array, a white cloud on its way from Cuba to Puerto Rico. And then the first waves of headaches would come, your brain calling out its last orders for oxygen, cramping like a muscle that's been pushed too far. That might give way to dizziness or to light-headedness or to sickness, but all of that unpleasantness would eventually pass. In time, you would forget why you had been sweating. It would all begin to seem like a movie, as though you were watching someone else's nightmare, as though the distance between you and the station wasn't anything at all, as though you could swim on over and climb inside and crawl into your sleeping bag, if only you wanted to, if only you weren't so tired. If only you cared . . .

Your training might kick in again, one last gasp, and you would try to shake off the dreaming, but your fight would wane along with your ability to concentrate. The brain has a built-in kindness, a genetic predisposition toward self-mercy: it goes first. It might still fire enough to register the tingling in your hands and your feet, the dryness in your throat, the little earthquakes that seized random muscles in your arms and legs and chest, but mostly it would be occupied with snuffing out awareness and replacing reality with good feeling. Close to dying, your brain would fill with euphoria, one final, blissful push into the ether.

The hallucinations would pile one on top of another, the whole of your life, real enough for you to see and hear and touch, burning up the very last of your oxygen. And then all of it at once would fade into black. Your eyes would remain open, your body stiff, but your brain would have finished signing off, catatonic, waiting only for the rest of your organs to follow. One by one they'd pack up and join the parade out of town: kidneys, liver, pancreas, spleen, lungs, and, finally, your heart. That would be the end of it. That would be the death of an astronaut, like drowning without the struggle, a man left empty instead of filled up.

That is, if you're one of the lucky ones.

. . .

If you're unlucky, you would come out on the short end of the 1-in-496 odds that you're on the flight path of something moving really, really fast. There are a million things it might be: a micrometeor, a hailstone, a piece of trash that slipped out of Mir and never burned away, even something from station. It might be a dropped tool, a bolt, the good luck charm that somehow slipped out of your pocket forty-five minutes ago. Then you would know the true meaning of orbital karma, first documented by John Glenn, who passed through his own exhaust every time he circled the earth in *Friendship 7* and wondered whether he was cutting through fireflies. That's when it was discovered that in space, what comes around really does go around, and too bad for you if you are in its way.

If it hit you somewhere that counted, in the head or the chest, you would die instantly, as if struck by a bullet going ten times faster than a gun could fire it on earth. If it was a large bullet, it might tear your head clean off, and the technicians in Houston would be left scrambling, wondering why your vital signs went from near-perfect to nil in the time it takes to sneeze.

But if it was something small, if it was the MADE IN JAPAN label that had peeled off some tin-can satellite or a wayward rock the size of a raspberry, and if it hit you somewhere other than your head or chest—if it caught you in the arm, say—well, then, that would represent some stone-cold misfortune. Because your brain wouldn't have nearly enough time to work its magic. Against the impending sensory overload, your life wouldn't have the chance to flash before your eyes. Instead of a seven-hour-long farewell, you would have nine horrific seconds to make your peace.

First you'd feel your skin break at the point of impact, and your bones shatter into chalk, and the rush of white blood cells to the hole that had been carved out of you. The pain of it would have pushed you into shutdown if the terror wasn't steaming in so close behind, because now, suddenly, the worst thing of all is that only the instant before, you had felt more alive than you ever had. You would know only too well what was happening, and what was about to happen, and that the hole in your suit was graver than the one in you. You would recall instantly the cold language of the space medical experts whose research you had read, like that of the American College of Surgeons describing the effects of and countermeasures to explosive decompression: "These insults are likely to be lethal, precluding the requirement for medical care."

Those insults would include, first and foremost, pulmonary barotrauma: the vacuum of space would empty out your lungs with a loud pop. Some of the air would come out of your nose and mouth with enough force to blow out your sinuses or dislodge your teeth. More of it would tear through the walls of your lungs, left flapping against the outward rush like tissue paper, and fill your thoracic cage. The rest of it would be pushed into your bloodstream, great bubbles of air suddenly choking your body to death. Those bubbles would enter your joints and paralyze you stiff, like the bends that you had worked so hard to avoid. They would also clog your veins shut, which would stop the flow of blood through your arteries, which would cause your heart to quit. Never mind, then, that every drop of water in every last one of your cells would still be turning into vapor, expanding your body to twice its normal size, squeezing your eyes out of their sockets, stretching your skin to its splitting point, and turning your ear canals into oceans. Never mind that all of that water, depending on whether you were in the sun or the shade, would begin either to boil or freeze almost instantly. Never mind that the gases trapped in your stomach would explode it, blasting your diaphragm upward and crushing the last scraps of your lungs, or that your large and small intestines would have been left in better shape had you swallowed a hand grenade. And never mind that all of the galactic radiation in the universe would pour into your body's new openings, cooking you from the inside out, if only your insides weren't already outside and your outside hadn't been blown to bits. Never mind all that.

Only those first nine horrific seconds would really matter.

One Mississippi . . . two Mississippi . . . three . . .

. . .

Jump.

Having clamped shut the worst parts of their imaginations, Pettit and Bowersox made the leap outside, each hooked on to his own length of steel cable. They were careful not to cross each other's paths or tie their lifelines in knots. They were also careful not to look down as much as they might have liked. There was work to be done.

All told, astronauts will perform 160 space walks to finish building their home, during which more than a hundred separate components will be fitted, wired, plumbed, and bolted together. On this walk, only the second American space walk to originate entirely from within station instead of the shuttle (called a "staged EVA"), Bowersox and Pettit were to put the last touches on the P1 truss, the massive girder assembly that *Endeavour*'s crew had delivered along with Expedition Six.

Because it would eventually help funnel the juice generated by the station's football-field-size collection of solar arrays, the truss was wisely equipped with a radiator. For every watt of power it generated, a watt of heat would need to be released, which the hostility of space would help along by providing some mighty chilly shade. (There's a 400 degree difference between light and dark up there, which helps explain why galactic radiators have ammonia running through their veins instead of water.) But for all of its importance, the truss's radiator seemed a fragile bit of hardware. To protect it during the shuttle's launch and the truss's installation, it had been folded safely away and locked down tight. Now Bowersox and Pettit headed out to pick those locks.

Hanging on to the truss—while traveling 17,000 miles an hour—and watching the earth spinning so quickly beneath them, each of them was reminded of those vintage black-and-white photographs of New York City ironworkers balanced on girders fifty stories up, lunch boxes on their laps, downtown opening under their asses. For Bowersox and Pettit, the bottoms of their drops were less concrete, but every so often, they shared the sensation—the same sensation shared by just about every spacewalker, high-wire daredevil, and riveter—that they were already on their way down, trapped in the middle of a long fall. Only when they zeroed in on their work or on some star in the distance, the way seasick passengers are instructed to stare at the horizon, did the feeling pass, left to percolate just under the surface.

It took them a while, in fact, to push through it fully, and by the time they had unlocked the radiator and prepared it for its slow release, Bowersox, Pettit, and their perch were about to pass through the day-night terminator. Because they needed to watch the radiator unfurl in daylight, they were told by the ground just to sit tight for the forty minutes it would take for the sun to shine on them again.

Pettit was more than pleased with the assignment. He delicately picked his way to the front side of the truss, which was still without power and unlit. There, the rest of the station's exterior lights were also blocked from view. Finally, he turned off his helmet's spot- and floodlights, and he was cast in nearly perfect darkness. He stayed there, hunkered down, and rode the leading edge of the night, looking up at the stars. He had never seen them so clearly. Without light pollution, atmosphere, or the distortion of a multipaned window, he saw colors that made him wonder if he had been dreaming after all. Not only did the stars not twinkle, but only a few of them gleamed white. Instead, they were intense reds, greens, and yellows, this great beautiful collection of pristine starlight, unfiltered and unhurried.

All too soon, however, the sun broke over the horizon, the stars were washed away, and Pettit had to get back to work. He and Bowersox set the radiator loose, and everything went just as the engineers on the ground had hoped it might.

Next, the two spacewalkers bounced their way toward the earth-facing docking port on Unity, which had a small amount of grit smeared along its edge. Houston's technicians feared that the dirt might eventually erode the port's seal and cost NASA a small fortune in oxygen loss, and they asked Bowersox and Pettit if they wouldn't mind cleaning things up.

It wasn't quite as simple as that. If the men used the designated foot- and handholds, the hatch remained just out of reach. They had to go for broke. Relying on their own ingenuity, as well as their gigantic brass balls, Bowersox locked his feet in tight, grabbed Pettit by his legs, and cantilevered him through open space, like a trapeze swinger. Pettit grabbed hold of the lip of the hatch, cleaned away the grit (by dabbing it with the sticky side of a length of tape), and waited for Bowersox to yank him back down. It was exactly as scary as it sounds, but neither man allowed himself to think much about the maneuver before he did it, probably because it was the sort of stunt that a purely logical man would refuse to do. Bowersox and Pettit were each in their own worlds, however, and they saw only a job that needed to be done, and next they saw a way to do it, and then they did it without so much as stopping to catch their breath. Their faces lost behind their gold visors, their breathing mechanical, their movements stiff, they had become robots, machine-like and unfeeling except when they paused to take in the view.

Another job done, they looked down at their to-do lists, checked off their housecleaning, and found that they had only one more task assigned. They were as sad as they were relieved.

At first glance, it seemed easy enough: they had only to mount a light fixture on one of the station's older installations, the S1 truss. Like the radiator, the light—on the end of a stanchion about three feet long—had been locked down flat for safekeeping. Bowersox and Pettit were to release the stanchion, which was held tight by two bolts, turn it perpendicular to the truss, and bolt it back down.

The men reached it without difficulty, and they were able to unbolt the stanchion, but for all of their pushing and pulling, they couldn't pry it from its safe place. Neither of them could budge it, and worse, neither of them could figure out why. And while both of them wanted to finish the job, to complete a perfect space walk, after nearly seven hours outside and now struggling against the clock and fatigue, they were told to pack things up and head for home, cutting those loops of fabric on their way in. Not to worry, they were told. The next crew up would work it out.

It was a long push back to the hatch, a whisper of dejection having taken the ecstasy out of things. But after they had made it back to their front step, they were distracted from their small failure by a reminder of their early bad luck, when they had very nearly missed the chance to head outside altogether. Bowersox turned that loop of white fabric in his fingers before he cut it free.

"What are the odds of that?" he asked, unable to wrap his head around the math—how something so small could have turned into something so big.

"Well, maybe it happened to us because we were able to figure it out," Pettit said.

Bowersox had thought of another possible reason in the meantime. He pointed to the identification number on his spacesuit. "I didn't want to make a big deal out of it," he said, "but I'm wearing 3013."

The revelation caused both men to take light pause, lingering just a little in wonder. But there was more to their waiting. On their way to a hot dinner (thoughtfully prepared by Budarin), a sponge bath, and their approximation of bed, an exhausted Bowersox and Pettit nonetheless found themselves wishing that they could spend just a few more minutes outside. Through the thin glass of their helmets, the sunlight was brighter, the panorama was wider, even the earth seemed a deeper shade of blue beneath their feet. The pure oxygen pumped into their suits had helped boost the sensation: it was as if for all of their time on earth they had been half asleep, and now, finally, they knew what it meant to be awake.

. . .

Three months later, in early April, they would know that feeling all over again, a wish that they might have preferred to see go ungranted. One of the ammonia-filled connectors that helps the radiator keep things cool had somehow lost its thermal cover (a "bootie" in astronaut vernacular). The bootie protected things from the white-hot sun. Without it, the connector was at risk of overheating, slow-cooked like an ant under a magnifying glass. Eventually it would spring a leak, spewing gallons of ammonia into space until the radiator's stores were empty. The radiator would stop venting the heat generated along with the burning of every watt of electricity on station. Although the station's other radiators would probably compensate for the lost capacity, tests showed that several other booties risked coming loose, like the first. Were they to make their escape all at once, first the machine, and then the men inside of it, would begin to feel warm, and then hot, and then flash-fried. Bowersox, in his cool, collected way, decided that this was an undesirable series of events. He and Pettit had no choice but to head back outside, this time not to build their home but to save it—the fate of their beautiful ship and their tenuous existence inside of it now resting on the repair of a tiny, mundane part. Like the plastic owls that guard the shuttle from woodpeckers, these seemingly innocuous booties were life.

All over again, they prepared. All over again, they suited up. They checked their tether. They nodded. They emptied out the airlock. And they opened the hatch, this time easily, thanks to Bowersox's determined handiwork the last time around, flushed with the same thrill that had run through them before.

Except that this time, they came face-to-face with the night. This time, there was no sunlight to welcome them, nor was any shining off the moon. There were no bright stars, and there was no blue earth. There wasn't even a sense of up or down. This time, there was only the pitchest black, the kind of darkness that looks liquid, leaving even the stoutest man feeling as though it might swallow him whole. Bowersox and Pettit stopped for a moment, their hearts pounding in their chests, and together they stared into the deepest sort of abyss. As Nietzsche had foretold, it stared as deeply into them.

Before they had a chance to change their minds, Bowersox and Pettit took a deep breath and threw themselves off the back of their ship.

. . .

Because replacing the booties was a relatively small job—although an absolutely necessary one—the ground had come up with a long list of additional repairs for the two men to complete. They were also scheduled to be apart for some of their second space walk. Bowersox would fix the booties; Pettit had some patchwork electrical work to crack. Together, they would be alone, connected only by their voices in each other's headsets. Should their tethers snap, there would be no hand to reach out for them, no eyes to see the last horrified looks on their faces.

Bowersox strung his way across to the first bootie. It was still there, but the Velcro tabs that had kept it in place had come undone, and it was hanging by the proverbial thread. He tugged it back into place, pinched together the Velcro seal, and then, to make sure it wouldn't try to take flight again, he retrieved some Russian-made wire ties out of one of his pockets.

They were another of those simple, indelicate things that the Russians had come to rely on in their time in space. Now the Americans liked them, too. The ties were made of short lengths of soft copper wire with loops at each end—they looked as though they might have been cut from a coat hanger. But when something absolutely, positively, needed to stay tied down, they did the job. Bowersox wrapped one around the troublesome bootie and twisted its ends together, an ugly but perfectly sound repair, the sort of practical handyman's touch that holds together our billion-dollar machines after we've sent them into space. The other loose booties received similar attention.

Meanwhile, Pettit replaced a circuit breaker near the base of the station's robotic arm and rewired a gyroscope. After he was finished, he joined back up with Bowersox, who had completed his bootie patrol. Each didn't mind at all returning to the other's company. Together again, they made their way to Destiny, where some plumbing awaited.

Under a micrometeorite shield, which Bowersox and Pettit popped open like the hood of a car, they found the heat exchanger that connects the internal and external cooling loops. The outside's ammonia met with the inside's water at that junction, and it needed a quick touching up before some unhappy alchemy took place.

A child's dreams of space travel don't usually include such mundane tasks, but that spot weld was exactly the sort of grunt work that keeps ships afloat. Somebody has to scrape off the barnacles. And while not every plumber has a few hundred carrier landings under his belt or has sampled gas from New Zealand volcanoes, whether it's in space or under the kitchen sink, pipes spring leaks and somebody needs to fix them. It just so happens that on station, there are only astronauts to call upon.

. . .

There was one last job that Bowersox and Pettit had started and now both wanted desperately to finish. They wanted to swim over to the S1 truss and get even with the light stanchion that had foiled them three months ago. They were determined to release it from its bear trap. More specifically, they had set their hearts and minds on hitting it, really hard, with a hammer.

Choosing their weapon had betrayed, somewhat unexpectedly, yet another difference between the Russian and American attitudes toward their machines and how best to treat them, the common ground so recently found in copper wire excepted. Within the respective toolboxes were chisels, wrenches, screwdrivers, and philosophies.

Outside, strapped to the side of Unity, there is a large, soft-sided bag, about the size of a bar fridge, filled with Yankee hardware. Pettit had taken to calling it the Little Bag of Horrors. Every astronaut posted to station, the spacewalkers of Expedition Six included, had learned to dread having to open it. Behind its benign-seeming exterior, perhaps two dozen tools were hiding in wait, tied to the ends of various tethers, some of which kept better hold than others. Whenever the bag was opened, all manner of implements would come bobbing and weaving out, some anchored, some not, leaving the astronaut in front of it doomed to a mad tool-herding scramble. It was as though the men were floating just under the ocean's surface and had been charged with harvesting kelp in a stiff current. For every one or two strands they managed to gather, another one would slip away or bind up its neighbor, until the flummoxed spacewalkers were left facing down an almighty tangle.

Reluctant to make like jugglers, Bowersox and Pettit had elected to bring their tools with them from inside station. Before they had ventured outside, they had made another important executive decision: when it came to sorting out the light stanchion, they were going to go with the Russian hammer rather than the American one.

The Americans—not wanting their astronauts to go too much to town—had delivered a thin, delicate hammer, clean and pretty. It wouldn't have looked out of place hanging from a rock collector's hip or waiting on a watchmaker's lathe.

The Russians, meanwhile, had sent up something medieval. Their more practical technicians had learned that sometimes brute force is necessary; the universe, after all, was built on collisions. The Russian hammer, then, was an eight-pound sledge. Not only that, but the bottom of its handle had been turned into a kind of sharp-edged pickax. The Russian hammer had some serious teeth.

Now, with one hand holding the truss and with his sledge in the other, Pettit—even in weightlessness—could imagine the heft he wielded. He had hated leaving this job undone the last time around, and now he relished the chance to knock the stubbornness out of the stanchion. The ground warned him not to swing more than once every twenty seconds, leaving enough time to let the vibrations dissipate and keep the station's gyroscopes from going haywire. That being said, and with him having promised to abide, Pettit was given permission to wail.

He reared back for a big swing. The friction between his shoulder and his spacesuit didn't feel particularly good, and when he brought the hammer down, it felt as though he was swinging through water. It wasn't all that he had hoped for. The hammer bounced against the stanchion with a satisfying shudder—Pettit could only imagine the clang it would have made on the ground— but the blasted thing didn't budge.

Pettit waited for what felt like a long twenty seconds, adjusted his stance a little, and took another swing. Nothing.

Twenty more seconds, and again, he took his best shot. And again, the stanchion remained unmoved.

Again.

Again.

And again.

Until, on lucky swing number nine, with his lungs heaving and his shoulder rubbed raw, Pettit's hammer met with success. The stanchion popped loose, floating free. Smiling ear to ear, he grabbed at it and showed it off as though it were a champion steelhead. He had won.

Triumphantly, he and Bowersox set about bolting the stanchion back into place. As they did, they noticed that in spots along its length, this precision piece of engineering had been dinged and dented, and they joked about Pettit's contribution to its final design. He delighted in thinking that in a decade or two, spacewalkers on the truss might look at the light and marvel at how often it had been struck by space junk. They would think it a magnet. But in reality, in those dents, a rookie astronaut named Don Pettit had left behind a tagger's graffiti, the way drywallers sometimes sign the back of a slab, and when their wall is torn down, as they almost inevitably are, the demolition workers will know at least who had laid it. In that way, Pettit and his Russian hammer had guaranteed their place in the International Space Station's tactile history. Together, they had made their mark.

· · ·

Bowersox and Pettit had tied up the last of their loose ends. Happily, sadly, they strung their way back to the airlock. Outside the hatch and its sliced-up fabric loops, they stopped to take one last look at the infinite space that opened just inches from their faces. It was beautiful. They tried to take mental snapshots of these fleeting, misty-eyed glimpses, each of them wondering whether he would ever have the chance to touch the universe again.

After making their last goodbyes, they ducked their way into the airlock and shut the door behind them. It felt as though they had come into the warmth from a blizzard without having to brush the snow from their shoulders.

Safely locked away, they came down. Their adrenaline supplies were tapped out, and they felt heavy with exhaustion, as if it was all they could do to strip off their gear. But in the hours it took them to undress and adjust their bodies back to life's normal pressures, Pettit was kept awake by a puzzle, by hints of a smell that he couldn't quite place. It had come inside with them, embedded itself in the white fabric of their suits. And now something about it had latched inside of him.

It was metallic, but it was more than that. It was sweet and pleasant. It was the smell of space.

If something had gone wrong out there, had one of those rubber seals split open or a 1-in-496 long shot come through, that smell would have been the last piece of data left for his brain to collect. Now he breathed it in again, then again. For some reason, the smell reminded him of summer.

And there it was.

During college, when Pettit had spent his vacations repairing heavy equipment for a small logging outfit in Oregon, he'd fired up an arc-welding torch to do it, and that torch had given off a sweet, pleasant, metallic smell.

Here was that smell all over again. And just as suddenly, space smelled for him like summer, the same summer whose arrival he'd watched through his window, greening earth's landscape without him.

9 MISSION CONTROL

In their prophetic novel, *The Return*, Buzz Aldrin and John Barnes sketch out a scenario that is one part nightmare and one part dream.

During another one of their bust-ups over Kashmir, Pakistan and India take their war to the skies. Hoping to knock out India's network of spy satellites, Pakistan launches a patched-together proton bomb into space. Too bad they overshoot their mark and manage to irradiate every scrap of hardware in orbit, including the International Space Station. On board, alarms sound, and radiation levels spike so high that they threaten to kill the three resident astronauts within hours. They are ordered by the ground to evacuate in the *Soyuz* capsule glued to their hull, the lifeboat that, in reality, makes like an escape pod straight out of hammy science fiction, the bucket of bolts that somehow reaches hyperspace. But when the astronauts power up the capsule by remote control, its thrusters inadvertently fire, burning off its fuel stores. Before the crew even has the chance to wedge themselves inside, they are flat out of gas. And because of an earlier fatal accident—on *Columbia* no less—the shuttle fleet is grounded. Besides, even if a shuttle could be readied in time for an emergency launch, the radiation levels in space remain too high for safe rescue. All of which leaves Mission Control with a dilemma it hasn't faced down since *Apollo 13* nearly self-destructed en route to the moon: How do we get our folks back home?

Good thing for the glowing, fictional astronauts—and not coincidentally, perhaps, given Aldrin's real-life entrepreneurial interests—private companies have begun developing their own rockets and are standing by. One, in particular, is primed and ready for its first flight: the *StarRescue*. Not only can it be readied faster than the outdated shuttle but it can also be lined inside with giant bags of water to prevent its crack crew from turning into mutants on their return.

Out of options, NASA's embattled honchos reluctantly agree to subcontract this life-and-death mission, and the *StarRescue* rockets into the black. Its crew approaches station, lassos the sparkless *Soyuz*, yanks it out of the way, and docks in its place. Once inside, they find the ailing astronauts hanging on to their last traces of life. Their hair is falling out. Blood runs out of the open sores that pockmark their emaciated bodies. There are gaps in their smiles where they once had teeth. Not a moment too soon, they are zipped into pressure suits and floated into *StarRescue*, which undocks, drops out of orbit, and finally makes a flawless landing on the dry lakebed at Edwards Air Force Base. First ambulances and next the president pull up beside the spaceship, shining white in the desert sun. It's a triumphant scene. Everybody is safe. Everybody makes it home.

. . .

No such luck for Sean O'Keefe and company. The whispered conversations that Ken Bowersox and Don Pettit had imagined taking place back home did, in fact, start taking place throughout NASA's sprawling complexes. With no *StarRescue* in wait, O'Keefe instead called to order a series of brainstorming sessions, trying to find a solution for his dilemma. But even the wildest imaginations were of little help. Today, when it comes to getting men from earth to space and back again, there remain only so many options.

For Americans, of course, the space shuttle is it. Starting the day after *Columbia* broke apart, there were conversations at NASA about how quickly the fleet might return to flight. Perhaps it was worth the risks presented by lifting off on short notice—and before anyone could begin to fathom why *Columbia* had failed so terribly—to retrieve Bowersox, Budarin, and Pettit. But that scenario raised two questions, the second more awful to contemplate than the first. Would Expedition Seven still replace Expedition Six, embarking on their own open-ended mission? And what would happen if the shuttle and crew that were hustled up in a hurry were vaporized, too?

Long before he sat down to listen to the grim forecasts of his advisers, O'Keefe had the last question answered. Already, NASA was taking hits. In the February 10, 2003, edition of *Time*—the issue that carried a cover photograph of fire and smoke over the headline THE COLUMBIA IS LOST—the agency's manned space program came under attack. In an article titled "The Space Shuttle Must Be Stopped," Gregg Easterbrook wrote: "The space shuttle is impressive in technical terms, but in financial terms and safety terms no project has done more harm to space exploration . . . This simply must be the end of the program." While he was at it, Easterbrook also took aim at the International Space Station. "There are no scientific experiments aboard the space station that could not be done far more cheaply on unmanned probes," he wrote. "The only space-station research that does require crew is 'life sciences,' or studying the human body's response to space. Space life science is useful but means astronauts are on the station mainly to take one another's pulse, a pretty marginal goal for such an astronomical price."

Easterbrook was not alone in his assertions. Across the country, there were calls for NASA to receive its termination notice, for its budgets to be slashed, for its dicky shuttle never to touch space again. Now, if another shuttle blew up—if, however heroically, NASA saw consecutive crews buried in flag-draped coffins—the agency would almost certainly lose its license to fly. For years, for decades, perhaps even forever, Americans would be grounded. There would be no more fire, no more experiments, no more giants. There would be only a more permanent gravity.

· · ·

With so much at stake, the idea of sparking a shuttle was dismissed almost as soon as it was raised. Next, a kind of Avdeyev Option was bandied about, although not in the way that Bowersox and Pettit had imagined it might have been. Because no one knew whether there could be an Expedition Seven, because Expedition Six remained physically fit and psychologically sound, and because, most of all, the three men had seemed so sincerely happy to stay, perhaps it was best to hold steady. Perhaps the trio could remain stashed away, safe so long as they didn't try to come or go, until the remaining shuttles had been checked out and the wounds left by *Columbia* had been allowed to heal. Perhaps it was best if everybody just laid low.

But like every answer, this one, too, raised only more questions. What if Expedition Six suddenly ran into trouble? What if some illness lurked in them that hadn't yet surfaced? What if the remaining shuttles were kept in hangars for as long as they had been after the *Challenger* disaster? How likely was another two-year-long hiatus? Could their bodies resist breakdown for such an epic stretch? Was it too much to ask for three husbands and fathers (and their wives and children) to remain apart for as long as Pettit's boys had even been alive? Who would call Micki and Annie and tell them that their men might miss two more birthdays, two more anniversaries, two more New Year's Eves? Or might they miss three? Or four?

Extending a mission was one thing. Extending it without end was another.

Sean O'Keefe shook his head.

He still didn't know the how of it. He still had no magic at his disposal, no potions or beans. There was no phone call he could make, no snap of his fingers, no name or face that he might recall that would make everything better. And yet his gut remained unmoved. Expedition Six needed to come home.

. . .

Soyuz. It was all that NASA had left. Once Americans had feared the tiny green capsule because of what it might one day do. Now they were left worrying about what it might not.

Unlike most rockets, *Soyuz* was born almost entirely of one man's inspiration: Sergei Korolev, the previously anonymous so-called "chief designer" of the entire Soviet space program.

Korolev was an engineer who flew homemade gliders and helped design military aircraft until 1938, when, at age thirty-two, he was swept up in one of Stalin's paranoid purges. He spent months packed into a boxcar on the Trans-Siberian Railroad, shuffled into a prison ship, pushed off at the Siberian port of Magadan, and sent down into the infamous Kolyma gold mines.

Later, his genius missed, he was transferred to a special prison outside Moscow where he was put to work on the war machine, mostly designing rockets for airplanes and missiles. After the war, he tore apart a few of Germany's V2 rockets until he was imprisoned again—although why and where remains in doubt. Following Stalin's death in 1953, Korolev was "rehabilitated" and elected to the Soviet Academy of Sciences. Always with his eyes and mind turned toward the skies, he poured his pirated knowledge and years of cell-bound imagination into the R-7 rocket, which was meant to carry a nuclear warhead far enough to reach New York City. The rocket proved poorly equipped for that unkind purpose, but following several explosive failures, it did prove a first-rate space booster. After Korolev convinced an embattled Nikita Khrushchev of the political value of launching an artificial satellite, he personally oversaw the development and delivery of *Sputnik* into space in 1957. (He lived in a small house in a grove of trees just a ten-minute walk from the Kazakh launch site.) Next came the launching of larger satellites, including the first one with a heartbeat, a dog named Laika, and in May 1960, a crash test dummy.

Despite his success, Korolev remained as invisible to the outside world as Tyuratam, erased from photographs the way the city had been from maps. Khrushchev wanted the Soviet success in space to remain one of the people, of the nation, and not that of a single brilliant man. Korolev was forbidden to travel or communicate with rocket scientists outside the Soviet Union. Even behind the Iron Curtain, he became a sort of omnipotent apparition, almost godlike in his status but equally unseen.

He was also fallible. A little more than four months after he had successfully launched his mannequin, his program met with an almost unimaginable disaster, the so-called Nedelin Catastrophe. In October 1960, an unmanned moon-shot rocket exploded on the launchpad. Among the casualties that night was Field Marshal Mitrofan Nedelin, the top Soviet missile general. Under political pressure and time constraints, he had ordered an inspection of the rocket—which had failed to ignite moments earlier—without first removing its payload of fuel or waiting for morning. Appearing out of the dark, as many as three hundred scientists, engineers, and technicians began crawling on and around the monster, looking for a cure to what ailed it, when its engines ignited and exploded. Everyone on the pad—including Nedelin, but not the normally hands-on Korolev—was killed, some by sonic shock but more by the fireball that lit up the flats.

It took six months for Korolev and his program to recover, but it did so in spectacular fashion: on April 12, 1961, Yuri Gagarin, poised to become the first man in space, lifted off in the rocket *Vostok*. It was not beautiful but it did the job, and it did it first. It also bore certain attributes that would later become part of the mainstay *Soyuz*. Gagarin was strapped into a capsule that touched down on land (unlike the American splashdowns), under parachutes. Also in contrast to its American counterparts, *Vostok*'s function was almost entirely automated; there were fears that even the best cosmonauts would be rendered incapacitated by weightlessness and stress. Although those fears proved unfounded, the Soviets elected to remain faithful to their totalitarian machines. Each of their subsequent successes—longer-duration flights, twin launches on parallel orbits, and the safe return of Valentina Tereshkova, the first woman in space—was built on the premise that the fewer buttons the pilots had to push, the less chance they had of blowing themselves up.

The Vostok program ended in the spring of 1963; next came the cobbled together *Voskhod*, a stopgap to fill the wait until the new *Soyuz* was ready. Nikita Khrushchev had demanded that Korolev counter the American Gemini program and its two-man crews (until then, rockets on both sides had been designed to carry a single passenger) with a vehicle that could host three. And while *Soyuz* was already deep into its planning stages, so many dreams turned into sketches and blueprints, it would never be hatched in time. Instead, *Vostok* was stripped down and emptied out, and enough room was found for three cramped men, with only enough supplies for single-day missions.

Necessity gave birth to an invention that the Russians have trusted ever since: because there wasn't room inside the capsule for ejection seats or reserve parachutes (*Vostok*'s pilots ejected minutes before the capsule landed and floated down to earth separately), a small solid-fuel impact rocket was added to the design. It was fused to fire just seconds before touchdown to make the dry landings feel less like the crew's space elevator had hit the bottom of the shaft. In the end, there were just two touch-and-go *Voskhod* flights, allowing the Soviets to record two more space firsts—the first multi-man crew and the first space walk—but it wasn't well-designed or safe enough, even by the Soviet definition of the word, to keep flying. Development of *Soyuz* and its accompanying proton booster was picked up again, but not soon enough for Korolev to see it fly. As a final indignity, he died on the operating table in 1966, moments after a hatchet job on a bad case of hemorrhoids.

Two unmanned *Soyuz* capsules were launched successfully the following winter. And then, on April 23, 1967, came *Soyuz 1*, piloted by Colonel Vladimir Komarov, the first cosmonaut to visit space twice. He had survived *Voskhod*'s inaugural manned flight, but he did not live through a second maiden voyage. During his descent, his parachute lines tangled, and his capsule crashed into the Kazakh steppes. Although the reasons for the tangle have never been made public, its terrible effects were: Komarov was killed instantly, having broken his hips, back, and neck, and rupturing virtually all of his organs.

Still, the Soviets remained wedded to Korolev's original design, and over the coming years—after several successful launches, albeit marred by the occasional nonfatal malfunction—*Soyuz* proved a simple, if inelegant, workhorse. It wasn't until the loss of the Salyut 1 crew in 1971, when three heroic space-station celebrities were suffocated and laid out in the grass, that the first in a series of redesigns was undertaken.

The Salyut 1 tragedy had revealed a fundamental flaw in the *Soyuz*'s original design, one that had plagued it since Khrushchev had made his demand for a capsule that could fit three men instead of the usual one or two. For Korolev to make that possible, he had been denied the room for even the most rudimentary safety systems or, as incredible as it might seem today, for the cosmonauts inside to wear spacesuits. In the sad case of Salyut 1's lost crew, the faulty valve that did them in could have been closed manually (and there was evidence that the crew had tried to shut it in their desperate last seconds), but the process took at least two minutes. It took half that time for *Soyuz* to empty itself of air. To have given the three doomed men any chance of survival, they would have needed their own supplies of oxygen, pumped into helmets. In the year of modifications that followed, it was decided that two-man crews were the better bet, and room for oxygen tanks and spacesuits was found.

Still, *Soyuz* was not perfect by any stretch. Since its first tangled parachute, there has been a weirdness in the machine, an almost natural predisposition toward glitches and snafus—most of them harmless and perhaps even endearing, but downright demonic every now and then. Although *Soyuz* hasn't carried corpses since 1971, it has come uncomfortably close to reaching the status of tomb several times. In one instance—*Soyuz 23*, which flew in October 1976—it landed off course and splashed down in near-freezing Lake Tengiz. The capsule's inside temperature plummeted, and its occupants were forced to wait several hours until recovery crews could connect a line and drag the vessel and the numb men inside to shore and safety.

But that experience was benign compared to the almost comically bad-luck flight that would have been *Soyuz 18*.

On April 5, 1975, Vasily Lazarev and Oleg Makarov first ran into trouble some ninety miles up, when their rocket's spent third stage held on longer than it should have—probably because the exploding bolts meant to help jettison it didn't fire—pushing them into a violent tumble. (The tumble rate was so far off the scale that technicians on the ground didn't believe the data; they ignored the problem until they heard the crew swearing loudly over their radio.) After their rocket's fat ass was finally ditched, Lazarev and Makarov's emergency reentry exposed them to g-forces that *Soyuz* had never been designed to handle. The spacecraft's theoretical limit was 15.0. When the weight rocketed past 18.0, the meter broke, and the cosmonauts each topped out at a ton and a half, snapping their ribs. Thankfully their parachutes opened and broke that hard drop, and their capsule made a relatively gentle landing on a snow-covered mountainside. But like a sled, it began speeding down the slope before the crew had managed to clamber out—narrowly avoiding falling over a cliff, like a barrel dropping over Niagara Falls, when its lines snagged on some pine trees.

The last remaining concern for the cosmonauts was that, after traveling 2,000 miles in just fifteen minutes, they might have landed themselves in China and thus, likely, in a Peking prison. Fortunately, they had landed just short of the border. They learned as much when their rescue team showed up, in the form of a band of Russian villagers who had watched in wonder when these spacemen fell out of the dusk.

. . .

Since that harrowing night, *Soyuz* has been tinkered with, usually without enthusiasm, with the exception of one significant overhaul, which saw its capacity increase from two men back to three. Still, its fundamental architecture has stayed the same: it is a child of the 1960s and a flying tank, ugly and hard-edged.

Like the *Apollo* capsules of old, it is launched on top of a booster rocket that dwarfs it in size. After it has reached space, the *Soyuz* capsule separates from the rocket and takes on a more manageable scope, leaving its crew with just nine cubic meters of living space spread across three small modules. The first, the orbital module, is vaguely spherical, capped with the docking mechanism that allows it to mate with its designated port after reaching the International Space Station. (For launch, it is usually filled with cargo.) The bell-shaped second module, called the descent module, contains the crew's three canvas cradles as well as banks of monitors and control panels, filled with rectangular plastic buttons marked in Cyrillic, like an Aeroflot cockpit. The third module—called the propulsion module, cylindrical in shape—houses the main engine, fuel supplies, and electrical systems, powered by two winglike solar panels.

Like every Russian rocket in history, it remains grimly functional and largely automated (although its crew has plenty of monitoring and switch-throwing to do). And upon its return to earth, during which the orbital and propulsion modules are ditched, it is reduced to that single bell-shaped descent module, saved from gravity by wind resistance alone. It is a barebones solution to the problem of launching men into space and returning them to earth. It resembles, in a lot of ways, those inventions that budding engineers come up with when they're asked to insure an egg that's thrown from a campus rooftop. Usually, the student engineers come up with a padded box attached to a parachute. What the Russians had come up with is *Soyuz*.

And now, it was all that NASA had come up with, too. For Sean O'Keefe, Bill Readdy, and the rest of the agency's upper management, it was not the happiest remedy. If Expedition Six really did drop into the *Soyuz* that was latched to the side of the International Space Station, Ken Bowersox and Don Pettit stood to become the first Americans ever to return to earth on a foreign vessel, the first Americans to stake their lives on another country's unhandsome technology. They would also become the first Americans since 1975 to come home in a capsule. It had been nearly thirty years since the last *Apollo* had splashed down into the South Pacific. All American astronauts since had glided back to earth, as in their dreams. Now it looked as though Bowersox and Pettit were about to be asked to fall.

More alarmingly, perhaps, the particular *Soyuz* capsule locked to station—*Soyuz TMA-1*—was the first in a new production run, replacing the archaic-seeming *TM*. Responding to the urging of the Americans, the Russians had included new instrumentation, monitors, and computer systems in their most recent effort. To their partner's chagrin, however, the Russians had never bothered to test-fly the new capsule. Its launch to station, in October 2002, had been its inaugural flight; its return would be its first descent, and there was always a chance that bugs had hatched in the meantime.

• • •

But aside from their bruised pride and a burgeoning case of the gulps, O'Keefe and company had a larger ill to contend with. Their first priority had to remain the safe return of Expedition Six. A close second was keeping the International Space Station operational, but they all knew that in rescuing the men, they now risked scuttling the machine.

Without the shuttle, the hopes of sending up an Expedition Seven had waned in the weeks after *Columbia*'s loss. It was assumed that, like the last crew of Skylab, Expedition Six would probably have to push station into a state of hibernation, dimming the lights and closing the doors, one step short of abandoning ship. Even that simple scenario presented its problems, however.

There was always the risk that the next crew to arrive on station's doorstep—not until the shuttle fleet was back up and running, not for two or three years at least—would fail to find their destination in the blackness of space. They might even push station out of orbit in their repeated stabs at it, and a multibillion-dollar enterprise would be lost, either cast adrift toward Pluto or burned into ash that would settle across an enormous blue sea.

More to the point, and contrary to the opinion of their critics, the International Space Station's crews did more than live out idle days in space, checking one another's pulse. They cleaned and maintained the place, unclogging filters and replacing booties, and without its live-aboard troubleshooters, station risked falling out of the sky long before the next men up would be able to shock its heart back to life. Experience showed that Mir's long tenure was made possible only by the creative and sometimes patchwork repairs undertaken by its crews. And although most of the new station's operating systems were controlled from the ground, there was no substitute for having someone like Don Pettit on board, merrily fixing a broken part or tightening some loose bolt. Leaving station empty also meant leaving it untended, and like a lakeside cottage locked up for winter, there was always the danger of a fire sparking, the roof sagging under snow, or a break-in by bears. In short, bringing home Expedition Six without first sending up Expedition Seven represented the sort of gamble that NASA's actuaries couldn't abide.

Not surprisingly, those same men still didn't care much for the idea of relying on *Soyuz* and *Soyuz* alone.

Sean O'Keefe had learned in his first days with NASA that within its walls, redundancy has never been a four-letter word. He learned that the agency's engineers and technicians liked to prepare for every imaginable contingency, mapping out giant decision trees on dry-erase boards—if x then y, and if y then z. But they also preferred it when there was more than one y and more than one z. Truth be told, they liked it best when they had the entire alphabet at their disposal. They liked their backup systems to have backup systems, and they were among the few paranoid people on earth who knew what came after tertiary in the sequence of orders. (Quaternary, quinary, senary, septenary . . . The most pessimistic among them really weren't happy until they'd reached something like duodenary, and even then, they slept with one eye open.)

No wonder, then, that whenever O'Keefe sat down with his advisers and their advisers and the advisers who were waiting in the wings after them, trying to come up with a fix, he was never left thinking that tonight was the night that he might get some sleep. With flowcharts and bullet points, the downbeats outlined every last thing that might go wrong with *Soyuz TMA-1*. It was a long list.

Perhaps in the months that it had been latched to station, the capsule had been struck by a piece of space junk, compromising its hull or its operating system. The new software that had been loaded into the *Soyuz TMA-1*'s main computer shortly before its launch might contain a glitch that had not yet been discovered. Perhaps all of its gas had leaked out into space in a fine, unseen mist, leaving its tank dry and its engines lifeless. The latches that had kept it tied to station might refuse to unlock. The shield designed to protect the descent capsule and its astronauts from the heat of reentry might have sustained a hairline crack during its flight up, which would turn it into a blast furnace on its way back down. The rockets meant to push the capsule into the earth's atmosphere might fire too early (dropping it into the Caspian Sea) or too late (dropping it into the Siberian winter) or not at all (dropping its three passengers into memorial books). Like the space shuttle, a sheared bolt, a loose electrical connection, an imperfect seal, a jammed valve, a ribbon of fatigued metal, or a faulty weld could turn the capsule into a death-trap. Unlike the shuttle, a hole in its parachute or another set of tangled lines could, too.

It always took the advisers much less time to run down the vaccinations they had should any of these nasty scenarios come to pass: none. Best among the worst case scenarios, Bowersox, Budarin, and Pettit might bundle into the ship, press the big black button that sparked its automatic reentry, and . . . nothing. That would leave them stranded—and then what?—but at least alive. Worst of the worst, something would go catastrophically wrong somewhere between up there and down here. That would mean the certain end of Expedition Six, and the probable end of everything else that was meant to come after.

Oh, and don't forget, we need to get somebody on station. We can't abandon it. We can't leave it empty. We'll lose it if we do.

Invariably these meetings ended with a long silence, a few more worry lines, and O'Keefe badly needing a cigarette.

. . .

And yet, trapped in their dead-end alley, NASA's brain trust finally found inspiration—in some ways, aided by the absence of choice. The questions they faced might have been painfully complex, but they knew that their answers had to be simple. The engineers didn't have what they needed to make themselves into architects. They knew that they had to settle for becoming machinists.

All right, let's start again: *Soyuz*.

True enough, it was all that O'Keefe and the rest had within their reach. If they were going to bring their three men down, they would have to take that gamble.

But it was also true that, since the beginning of station, *Soyuz* capsules had made the journey from earth to station and back every six months. These were the so-called taxi missions. Because *Soyuz* capsules were prone to a bad case of the yips when they were looked at sideways—let alone when they were bathed in the universe's metallic exhaust for weeks on end—they were exchanged regularly, every 180 days or so. The three-man crew who had flown *Soyuz TMA-1* to station not long before Expedition Six's arrival had come home on the *Soyuz TM-34,* which had remained faithfully latched in place, just in case.

Following this routine, it had been decided long before *Columbia* came apart that a new *Soyuz—Soyuz TMA-2*—would make the trip up at the end of April 2003, and the old one, *TMA-1,* would come down. However, instead of being manned by a courier crew, as had been custom, perhaps some sort of Expedition Seven could catch a lift up with *TMA-2,* and Expedition Six could ride shotgun home on *TMA-1,* already nearing the end of its orbital life span. The next time the *Soyuz* demanded switching out six months hence, another Expedition would arrive with its replacement, Eight for Seven, Nine for Eight, until the shuttles were ready to fly again.

Now all that remained was the age-old question of supply and demand. Even with their best conservation efforts, Expedition Six had confirmed suspicions that a three-man crew could not be sustained by *Progress* alone. Their food and water would run out, and NASA did not wish to bear witness to the first orbital famines and droughts.

Bowersox, Budarin, and Pettit were asked to inventory every little thing that remained on board station. Next, NASA's best math geeks calculated what patched-together crew could stretch out what patched-together grub allowance. Water was their biggest hitch. But with some prideful fanfare, the geeks announced that they could make a two-man crew work—so long as they weren't especially big eaters. By the time they had come to the last line of their homework, they had found enough room for one little American, one little Russian, and a whole lot of space in between.

O'Keefe took all of it in, each of the whisker-close projections and extrapolations, and realized that working at NASA had started to change him. For the first time in his budget-conscious life, he wished that he had more fat on the bone. He wished that he had more time to mull things over, to see if some magical cure might surface, and he wished that the absence of options didn't leave him feeling so boxed in. As a boss, he would have liked to have had some kind of choice to make. But in the end, one of the biggest decisions of his professional career wasn't much of a decision at all. Really, it didn't even amount to accept or reject. Everybody on NASA's food chain, starting with him, knew that there was only this one way out, and even it might be closed to them if they waited any longer to take a run at it.

No matter how badly it sat in their stomachs—no matter how strongly they disliked the idea of ceding control of their fates to the Russians and the lives of three astronauts to the stars—everybody knew, when they looked at O'Keefe, that he would have to nod his head. There was no other move for him to make.

. . .

The Russians signed off on the plan with big-chested pride. Expedition Six would come home on *Soyuz TMA-1*; Expedition Seven would go up on *Soyuz TMA-2*. Now the International Space Station was nearly as much a Russian enterprise as Mir had been. Now they would be solely responsible for shipping crews in and out and for supplying them, too; when it came to keeping station in space, the Russians were it. They would have the chance to prove, once again, their mean-edged superiority, at least when it came down to the hard work of having men locked into orbit. And right along with them, their freakish, beloved, donkey cart *Soyuz* was about to come out of the shadows and into the sun.

Bill Readdy rang up Expedition Six and told them that they would be coming home, and that a scaled-down Expedition Seven would take their place, and that both crews would make their trips in *Soyuz* capsules. Bowersox, Budarin, and Pettit were disappointed not to have been given permission to stay, but they were not surprised. Along with the geeks on the ground, they had done the math. (Already, just days after *Columbia*'s loss, Budarin had nodded at their hospital-green *TMA-1* and winked.) They accepted their latest orders in the way the men on the ground had issued them—with shrugs, resignation, and a crackle of nerves. While they might have liked for their new *Soyuz* to have been better broken in, there was something about being asked to ride a Russian rocket that had instilled them with a Russian fatalism. This was how it was going to be.

All that remained was the culling from the candidate ranks of the two men who would become Expedition Seven. The designated American turned out to be a brainy, easygoing flight engineer named Ed Lu. He had made the trip to space twice before, docking once with Mir and once at the International Space Station. Both stays were relatively short, but during the second, he'd had time to conduct a space walk with Yuri Malenchenko, a former commander of Mir. (The two men connected power and communications cables to the newly installed Zvezda module; they also, thankfully, installed the toilet as well as the treadmill.) The space walk had gone so smoothly, and they had seemed such a perfect match, that Malenchenko was asked whether he might join Lu. He accepted. For NASA's managers, having had so much wrested out of their control and having become so tired of resorting to dice-rolling, the pairing of Lu and Malenchenko provided a kind of respite. In those two men, at least, there was nothing to worry about.

. . .

So long as Expedition Seven made it to station, that is. Lu crammed the usual nine months of *Soyuz* training into a few short weeks, spending his days in simulators and classrooms and his nights hitting the books. It was a demanding, exhausting routine, broken only by the occasional photo shoot and the elaborate preflight rituals that the Russians continued to observe.

On one cool, bright afternoon, dressed up in suits and ties for the cameras, Lu and Malenchenko visited Red Square and, with bells ringing in the distance, placed flowers along the Kremlin Wall at the plaques commemorating Yuri Gagarin and Sergei Korolev. Later they went to the Gagarin Museum in Star City and signed the cosmonaut book with a silver pen—Lu again in his suit, Malenchenko in his blue military best; they stopped for a round of pictures with their girlfriends (and Lu with his mother as well) in front of the museum exhibits; and last they conducted a final press conference the morning before their historic *Soyuz* launch. In their blue flight suits, the men sat outside in the shade and faced reporters.

"I feel very good about the flight," Lu said, adding that to help him get through it, he'd packed some family photographs that he would pin up in his private quarters on station, as well as some small, secret keepsakes. "Just some things to remind me of home," he said.

On April 25, 2003, Lu and Malenchenko underwent their final physical checkups and began their last preparations for flight. After being helped into their white-and-blue spacesuits—including old-fashioned pilot's caps, leaving them looking a little like the original leatherheads, and on Lu's upper right arm, the patch worn by *Columbia*'s lost crew—the two men walked through a gauntlet of applause and flashbulbs, waving awkwardly at the assembled crowd. They boarded the police-escorted bus that would take them to the *Soyuz* launchpad, where they ran through another crowd, stopping and turning on the ladder that climbed up toward their capsule, waving again at a gang of shouting photographers.

Finally, they were strapped in, the engine and four boosters fired, and their *Soyuz TMA-2* lifted into an overcast sky. Within seconds, they punched through the clouds. The emergency capsule on top of the rocket and its boosters were jettisoned, picking up incredible speed all the way. More than one hundred miles up, five minutes into its flight, the rocket's second stage dropped away. After four more minutes of powered flight, the capsule separated from the rocket's third and final stage and slipped safely into orbit. Everything had gone perfectly. A crowd of anxious Americans on the ground—including Lu's fiancée, Christine Romero, and Jefferson Howell, who had delivered so much bad news to Expedition Six nearly three months ago—burst into applause. There were handshakes all around.

Romero, who had received one last kiss on the launchpad, was later interviewed by NASA TV and asked about her thoughts upon watching her man launch into space. Her answer was telling in more ways than one.

"First, when the shuttle," she began before stopping to correct herself. "Um, excuse me. When the *flight* lifted off, I think I was overwhelmed by the experience. I had no idea what to expect. Everybody says it's fabulous and it's brilliant and it's all those words, but it goes beyond that. I couldn't even speak. It was just so emotional, and I never expected to feel that way."

In that moment, she had learned an astronaut's hardest lesson, the same lesson that had been learned by Sean O'Keefe, Bill Readdy, Micki Pettit and Annie Bowersox, and those three men in a bucket. Between expectation and reality, between flying and falling— between earth and space, between home and away—there will forever remain some kind of gap.

In March 1965, the cosmonaut Alexei Leonov nearly died three times—first outside of his *Voskhod 2* capsule, then inside of it, and then outside of it again.

Leonov, under the watch of his crewmate, Pavel Belyayev, would be the first man to attempt a space walk. The Soviets had designed a primitive, inflatable airlock for their stripped-down capsule and a spacesuit that Leonov could fill up with air, like a deep-sea diver's, and attach to an umbilical cord that would prevent his drifting away. Then they ordered him to head outside.

Tumbling without anchor, Leonov watched in horror when his umbilical cord almost immediately twisted like a rubber band. In his struggle to work out the kinks, he lost track of time and found himself with only a few minutes to return to the relative safety of the capsule. To make matters worse, his suit had filled up with too much air in the meantime, inflated taut, and Leonov couldn't bend his legs enough to shimmy his way back through the hatch. He had become the original square peg.

Belyayev, trapped inside instead of out, was powerless to help him, able only to listen to Leonov's desperate voice over the radio: "I can't . . ." he said, his breathing getting fast and heavy. "I can't get in."

As a last resort, and at risk of suffocating himself, Leonov released most of the oxygen from his suit. That gave him the flexibility that he needed to squeeze into the airlock. It also left him with just enough air to exhale one big sigh of relief.

The following day, the crew was scheduled to return from orbit by firing braking rockets that would drop them into the upper reaches of the atmosphere. The time came to return. A button was pressed, and nothing happened. The rockets failed to light, leaving the cosmonauts trapped in orbit, where it looked as though they might stay forever, alive until they sucked away the last of their oxygen and then dead, a permanent fixture of the night sky.

Fortunately, a set of backup rockets belatedly fired. But the delay put the capsule 2,000 miles off course. They were headed for the rocky slopes of the Ural Mountains, where, they had thought, no one would be waiting for them.

It was late winter in rugged country, still and quiet in the deep freeze. After they had gathered their nerve, Leonov and Belyayev broke open their hatch. Their breath turned solid in the cold. Then the sun started setting, and their fingers and feet began to go numb in the last of the light. They dropped to the ground and built themselves a fire. The warmth of it was welcome; its glow, however, attracted a pack of wolves. Yellow-eyed through the trees and howling, they began circling the cosmonauts. The men might have liked to run, but they had the strength only to crawl—back into their capsule, where they shivered the night away, poking their heads out just often enough to take stock of their company, which proved reluctant to leave until troops arrived on skis the next morning.

Since that harrowing night, every Russian crew has packed a little something under the seat. In an otherwise empty space behind the small-man cradles in *Soyuz TMA-1*, there are soft-sided white bags that contained emergency supplies, including warm clothes, water, and—because sometimes bad things happen—a double-barreled sawed-off shotgun.

It had been waiting for Expedition Six all this time, packed away, bundled together with a small arsenal of cartridges and flares. Ken Bowersox, in particular, had taken account of it early during their stay. Had Don Pettit or Nikolai Budarin short-circuited with grief or homesickness and decided he wanted to reunite station and earth, all he needed was a few seconds and access to a hatch or a window. "In the event of that contingency," Bowersox might say, he had wanted a surefire way of convincing the potential escapee not to punch a hole through the side of the ship. He'd known all along exactly where that shotgun was.

Now, in the tiny spaces around the cannon, the three men began packing for their journey home. It was surreal almost, their slipping away, two months overdue and filled with mixed emotion: gathering up their things marked the end of the mystery, as well as the end of their waiting, but it also marked the end of their time in a place that had become magical for them. Sometimes endings can read like new beginnings, but for Expedition Six, it felt as though they had only those few blank pages that follow the epilogue left to turn through. This was it.

Only Budarin didn't seem saddened by the leaving noises. He was puffed up and excited, thrilled to show off to his friends the heart of the Russian space program. Also, the moment the three of them dropped through the hatch, Budarin would assume command from Bowersox. *Soyuz* would be his showcase, and in that short trip, he would jump from outsider to insider, from feeling like a guest to playing host.

Not that Bowersox minded. The prospect of flying in a new ship (or at least in a ship that was new to him) had brought out the old test pilot in him, and that was all that was getting him through. After riding in *Soyuz*, he would have a space résumé as complete as he had ever allowed himself to dream it might be: five dramatic space shuttle flights; a long, peaceful stint on the International Space Station; and now, a journey back to earth in *Soyuz*, the Russian vessel of legend. By the time he returned to Houston, Ken Bowersox would have every imaginary stripe on his shoulder and medal on his chest. The whispers that trailed him would start to sound like an ovation.

As for Don Pettit, it was one more surprise in an assignment filled with them. He had grown so accustomed to the extraordinary that it rarely made him blink anymore.

There was one hitch, however. In a capsule that once didn't leave room for its passengers to wear spacesuits, there weren't many places to stow luggage. A few critical experiments were bundled up and packed away. The emergency supplies were checked and rechecked. And now Bowersox, Pettit, and Budarin were left to decide what few personal effects they would carry home.

On the way up, in the relative vastness of the shuttle, they had been given small satchels, about the size of gym bags, that they could fill with books, photographs, music, and whatever earthly souvenirs they needed to get by. Limited by the cramped quarters of *Soyuz*, they were allowed to return with only three such mementos, all of which had to fit in a single, low-profile pocket on the leg of their Russian spacesuits. The rest of their belongings might one day follow them home, whenever the shuttle began flying again. But there was a chance that their treasures could be jettisoned like trash if station started looking too much like Mir.

Budarin wasn't weighed down at all by the dilemma. He was never one for sentimentality or affection for objects, and he just as soon would have stuffed his pocket with cookies and a second helping of Jellied Pike Perch (I). Bowersox didn't waste much time with his decisions, either. He packed up his favorite pair of blue shorts, the same pair that Pettit had soaked with that wayward sphere of orange juice, and a couple of golf shirts. The ugly tie stayed. Easy, done.

But Pettit was the only man in the seats at the auction in Los Alamos all over again, and this time with only a single pocket to fill. In the days before departure, he wrestled.

In the end, he decided that he would have to leave behind his beloved books and tools. (They might prove useful to subsequent crews, he reasoned.) The pragmatic engineer in him also decided to leave behind Micki's favorite necklace, which she'd dropped into his hands before he left and which he'd taken out and run through his fingers whenever he felt alone. He could always buy her another one. What he couldn't replace were two long-handled spoons out of the galley, designed for digging the dregs out of the bottoms of pouches, with holes punched into their ends so that they could be looped with idiot string and tied to his wrist, like mittens to a jacket. He had hated those strings, and against regulation he had cut them, but he had fallen in love with the spoons. Pettit thought they were beautiful in their shape and utility, perfect in their way. He imagined that he would give one to each of his boys, and they would take them camping, eating whatever they heated up over the fire right out of the tin without ever touching the sides.

So there were the spoons, one, two. His chopsticks made three.

. . .

Despite their packing up, it took time for Expedition Six to start really saying goodbye to the International Space Station, mostly because there wasn't yet anyone to hand over the keys to. Until Expedition Seven came in and began unpacking their own belongings, pinning up their own photographs, and listening to their own music, station was as much theirs as it had ever been. Their ship remained in their command and still felt every inch like their home. It was such a tough feeling to shake that, even while they were busy preparing their *Soyuz* capsule for launch, they felt as though they were in the middle of just another evacuation drill: *What would happen if we had to leave?* They had difficulty grasping that this wasn't more practice—that this wasn't make-believe, and theirs were not simple, isolated motions. There were consequences to them.

Soon this magical place for which they had such feelings would have strangers in it, oblivious to the time-honored routines and memories that had carried Expedition Six through almost six months of laughter and tears. Within a few days, all that would remain of their presence were bits and pieces packed away for safekeeping and the odd coffee splatter hidden away in a nook. They knew that. They knew that they were about to become ghosts.

Too soon, it seemed, Expedition Seven's own *Soyuz* began its final approach, one more in the long series of lights shining white in the distance. In just a little while, it took on its true shape. Its solar arrays came into focus, and its head and body became more defined. But still, despite its speed, there was no real sense of movement. From inside station, it looked almost fake, like a child's model hanging from a string.

Pettit began taking video of it, filming through the great window in Destiny. First he captured *Soyuz* against the backdrop of earth and, shortly thereafter, against deepest space. Cut together, the footage was almost artfully symbolic. Within a few frames, we see a long journey's beginning and end, as though the trip had taken place in seconds, having gone from blue to black just like that.

Radio signals bounced between the capsule and Zarya's docking port, the waves helping to act like a winch, drawing the two satellites together. Once they had drifted within forty meters of each other, Budarin talked Yuri Malenchenko the rest of the way in. *Soyuz* approached station at a relative crawl, two-tenths of a meter per second. Just as *Endeavour* had done nearly six months ago, the capsule slowed even more before contact, drawing out the anxiety of union. "If you're into docking mechanisms, it doesn't get any better than this," Pettit said, filming *Soyuz*'s long forward probe finally entering the docking port in a vaguely sexual hookup. A tremor concluded the embrace, the impact just strong enough to announce that capture had been successful. "Right on the money," was Houston's happy assessment. Everybody began breathing again.

More than an hour later, after a series of tests showed that there were no leaks in the seal, Budarin set about opening the white, cone-shaped hatch cover that separated Expedition Six from Seven. With Bowersox floating over his shoulder, watching closely, and Pettit taking photographs, now with his trusty Nikon, Budarin unlatched the hatch. Next, he took one, two, three tugs . . . without success. He kicked back, like a high diver lifting himself from the bottom of the pool, and took a wider view of his work, making sure that each of the latches was undone. They were, and he dropped down to try to open the hatch again. This time, with one good tug, it unstuck, popping like a plug being pulled out of a drain. It floated open to reveal a pair of smiling faces.

"Guys!" Budarin shouted in Russian.

Malenchenko was the first out, helped by Bowersox's extended hand. They hugged. Malenchenko then made way for Lu by falling into Budarin's arms.

"Hi, Yuri," Budarin said. "Congratulations."

Lu soon popped out through the hatch, hugged Bowersox,

hugged Budarin, and then gave a thumbs up to Pettit's camera. The five of them floated into Zvezda, with Pettit bringing up the rear and still clicking away, before they finally greeted one another properly and began talking to the ground in Russian and in English.

Among those beaming up good wishes was Bill Gerstenmaier, the ecstatic space station program manager. Having gone from nearly seeing station emptied to seeing it full, he let loose with emotion. "Sox, Don, Yuri, Nikolai, Ed, this is Gerst. It's great to see all of you guys on board the station. This is one of the happiest days of the program. To see all of you on board the International Space Station is just phenomenal. You guys enjoy your couple of days together and we look forward to Sox, Don, and Nikolai coming home."

"Thanks, Gerst," Bowersox said. "This is a real goldfish moment up here."

The expression "goldfish moment," was an affectionate wink across the miles toward Gerstenmaier. Expedition Six knew that his job, in a lot of ways, was thankless. When he wasn't trapped in yet another boring meeting in yet another poorly lit room, he was being besieged by one crisis or another—as though the best he could hope for, in asking for relief from the drudgery, was breathless panic. Only very occasionally was it broken instead by transcendent moments of beauty or grace. He called those "goldfish moments" after one of Gary Larson's more strangely poignant editions of *The Far Side*. In it, a band of murderous soldiers is storming into a castle over a drawbridge that crosses a moat, but one of their number has been stopped in his bloody tracks, distracted by something pretty in the water. "Oo! Goldfish, everyone! Goldfish!" he says.

With the nightmare of *Columbia* nearly over, and with the drama of the crew's return still days away, this was one of those rare, sweet moments set aside for watching goldfish.

It did not last long. For Bowersox, Budarin, and Pettit, almost doubling the station's population after six months of stasis made things feel a little tight around their shoulders. It wasn't so tight that they were unhappy for the company, but it was crowded enough to unsettle them, elbowing them into reality's harsh orbit. At last they began to understand that this was the end, and the knowledge knocked them off balance. For Bowersox, it felt as though he was being kicked out of his apartment for not paying the rent. It felt as though he was losing custody of something he loved.

"But it's great to be together," he said, trying to shake the unease. "It's great to see these guys, and it's great to be here on station."

The next day, April 29, the five residents held press conferences. Bowersox and Pettit were asked what they were most looking forward to, now that they were on their way home. For their audience on the ground, imagining these three tired, lonely men bundled into a crawl space, the hint of reticence in their voices was surprising.

"I'm actually going to miss station quite a lot," Bowersox said, admitting to a feeling that had been building in him over the past few days rather than subsiding, and one that he had debated leaving unspoken. "But when I get back to earth, the best part is going to be able to hug my wife and hug my kids."

Pettit followed. "Certainly being with your family," he said. "I've got two little boys who turned two early on this mission. They're talking sentences, and I've never been with them when they've been talking, so I'm really anxious to be with them and my wife. I'm looking forward to getting some good home cooking again, too."

A couple of days later, on May 2, their last full day in space, Expedition Six did two more ground conferences, these for *The Early Show* on CBS and with Miles O'Brien on CNN. It was clear that Bowersox was still struggling with saying goodbye.

"I'm going to miss flying, floating from place to place here in station, and I'm going to miss the spectacular views," he told O'Brien. "You just can't beat looking out the window and seeing our planet. We live on the most beautiful place probably in the universe."

It was unclear, just then, which home he was talking about. But now he consented to look ahead, if only for a little while. He talked about looking forward to feeling the wind blowing off Galveston Bay, and yet the sentiment rang hollow. It seemed as though he might never be able to turn his mind fully toward leaving.

But orders were orders, and after a fitful night's sleep, it was time to make the cleanest break possible. First came the change of command ceremony, a repeat, word for word, of the exchange that had taken place when Expedition Six took over from Expedition Five, only with Bowersox assuming a different role.

This time around, it took the combined crews three takes to record it cleanly. First, Ed Lu read from a prepared script, cold and clinical in a failed attempt to take the sting out of things: "Change of command of the ISS shall be an instantaneous transfer of total authority, responsibility, and accountability from one individual to another." He passed the radio to Bowersox.

"Today, I couldn't be prouder to be a member of the Expedition Six crew along with Nikolai and Don," he said. "Over the last five and a half months we've experienced some really sad moments and some extremely happy moments. But most important, we have managed to stay together as a crew.

"Ed, Yuri, you have to be the two luckiest guys who come from the planet earth today. Over the next six months, you get to live on board this beautiful ship . . . I wish you well, and I hope your expedition goes as wonderfully as ours has. I wish you many fantastic memories." Bowersox took a breath and said to his Russian replacement, "Yuri, I'm ready to be relieved."

"I relieve you of your command," Malenchenko said.

"I stand relieved," Bowersox replied.

And Pettit, standing behind them, rang the ship's bell for the last time. The note echoed through the cabin the way it had that hard day in February, the station and its residents having gone pin-drop quiet all over again.

· · ·

A jet-lagged Sean O'Keefe was patched through almost immediately, his Southern tenor's voice breaking the silence. He had made the long flight to Russia to watch the fireworks and shoot some vodka after his boys had made it safely home. He was joined by Bill Readdy and his blasted contingency handbook, as well as by a bright, fastidious man named Paul Pastorek. He was the head of NASA's legal team, but more than that, he had also been the best man at O'Keefe's wedding, and O'Keefe had been his. They had known each other for nearly thirty years, since college, and now, as always, Pastorek was O'Keefe's right-hand man.

When *Columbia* went down, the two men had been standing next to each other beside the runway in Florida. As soon as the touchdown clock had hit zero, Pastorek had begun madly taking notes, filling pages of his legal pads with the grim details of that terrible morning. Today, he hoped that if he found cause to start scribbling, he would be documenting a happier return to earth.

Still trying to shake his own memories of February, O'Keefe had watched Pettit ring the ship's bell and the crew fall silent and now struggled to sound casual, coming up with something uplifting to say:

"Congratulations to you, Ken, and to Don and Nikolai for a fantastic expedition mission. Y'all did a superb job. It was an honor to witness the change of command ceremony. Y'all followed the tradition with great honor as well as extraordinary capability. We appreciate very much all that y'all are doing and wish you Godspeed."

"Well, thanks," Bowersox said. "It only took us three takes. That's pretty good by Hollywood standards."

O'Keefe laughed. "Okay guys, put in your order for how you want your steaks done, so we can have 'em ready for you when you arrive."

"Oh, I won't need anything special to eat," the monkish Bowersox said, "but let me pass it to Don, maybe he's got an idea."

Pettit took the radio. "Medium rare is fine with me," he said before handing it off to Budarin.

"Rare," he said in his clipped English. "Two of them."

"Okay, gents, do well," O'Keefe said with another laugh. "Appreciate it very much."

"Thanks for talking to us today," Bowersox said. "We're going to get back to work here."

"Good to know," O'Keefe said, signing off.

The five temporary crewmates were in the middle of another round of hugs when the ground radioed back in: "That concludes the event. Congratulations to Expedition Six on a good flight, and we'll see you home shortly."

"Thanks Houston," Bowersox said. "It's tough to give up command of a wonderful ship like this. But it always comes, and you have to be ready for it."

Bowersox still wasn't. If he had learned anything in his time in space, it was how much living in that environment—the loneliness of it, the enchantment, the weightlessness, the views, the friendships built, and the memories made—had made him softer, more feeling. Highs were higher than they were on the ground, and now, as he had learned three months ago, he knew that the lows were lower, too. Just as he was about to fly again, about to do what he had dreamed of doing since he was a child in his father's front seat, his pilot's cool left him. And he had to admit finally, at least to himself, to the feeling that he had been afraid to give a name to, especially knowing that so many people were waiting so anxiously for him on the ground: in his heart of hearts, he didn't want to go home. He wanted to hide away in the quiet, sleeping against a wall of water, and waking up every morning to look through his window and see without fail sunrise or sunset, one always on the heels of the other.

But the choice was not his. Already he was reverting, returning to that lesser state he had known before he had rocketed to this beautiful place. Like an inmate shoved back into his cell after yard, he was practically pushed through the hatch and into *Soyuz*. Along with Budarin and Pettit, he folded himself inside a vehicle with a volume only a little larger than the interior size of a Dodge Neon. They might have liked for it to feel at least like a tank.

Expedition Six took one last look through the still-open hatch, back into station and at the two men who had taken their places. In that moment, the five of them were at the starts of their own incredible journeys, but three of them had a radically different destination coming into view. It was a hard farewell, and everybody's voices had gone quiet for it. They spoke to one another as though they had just finished throwing dirt on the casket.

"Bye, guys," Malenchenko said.

"Have a great trip," Lu said, snapping one last portrait when the hatch began to swing shut.

"Luck and success to you," Budarin said just before it closed tight, perhaps unsure who needed the blessing more.

. . .

Expedition Six pulled on their spacesuits inside the *Soyuz TMA-1*'s orbital module before floating down into the descent capsule and closing yet another hatch behind them. (Budarin, now in command, had the honor of severing last ties.) Bowersox, who would act as Budarin's copilot, took the left-hand seat. Pettit dropped into the one on the right. Budarin finally took the driver's seat in the center. Lying on their backs with their knees pulled up toward their chests, and with their legs already beginning to cramp, Expedition Six went through their preflight checklists, a series of procedures designed to make sure that their bubble wasn't going to burst the instant it began floating free. After they had finished, Budarin needed only to nod at Bowersox, who nodded back before the pair pressed the buttons and threw the switches that kicked off their return to earth. It was almost shocking to Bowersox how such a simple operation could have so much weight in it.

Exactly 160 days, 21 hours, and 50 minutes since they had last felt gravity's pull, Bowersox, Budarin, and Pettit heard the pins and latches that tied them to station unhinge, and they could feel the gentle spring ejection system that began to push the two vessels far enough apart to make it hard to remember that they had ever been one.

Unlike Expedition Five, who had watched their old haunt disappear through the shuttle's big windows, Expedition Six saw only the hatch that they had passed through grow smaller on the two monitors that had flickered to life in front of them. It was as though they had to say their last goodbyes through a filter.

"We are seeing separation," said the Russian ground at TsUP—Mission Control in Moscow. "Good luck guys, and a soft landing to you."

"Safe journeys, guys," Ed Lu said, watching their leaving, his hands pressed against glass. "Have a safe landing. We'll see you in six months."

"Good luck returning home," said the ground.

As with their launch, Expedition Six's first omens weren't everything they might have hoped for. It was crowded inside the capsule—Pettit actually had cargo stowed on his lap—and one of their cooling fans didn't work. They began feeling the first waves of claustrophobia. And they had hours to go, a goodbye stretched to the point of breaking.

On the monitors, Pettit watched the station's hatch continue to shrink. He marveled and smiled at how only minutes before, he had passed through it as easily as walking through a door, and now it was slammed shut to him and out of reach. The sudden gulf spurred him into thinking about the orders of home—house, street, neighborhood, city, state, country, continent, hemisphere, planet—and how the farther he had traveled, the farther he could be from his front door and still feel as though he had returned. When he went to the corner store, he wasn't home again until he was back on his couch. When he had jetted off to a foreign country, he had felt as though he was home as soon as someone stamped his passport. Now, he thought, he was well on his way to digging his two feet into solid ground, and in that instant, he would have made it all the way home, even though his front door would remain half a world away.

By the end of his dreaming, station had almost disappeared. It had looked at times as though it was pulling away from them, rather than the other way around, and in that moment, Bowersox, Budarin, and Pettit felt a little of what their wives had felt when they had first gone away. For the first time, they felt as though they were the ones who were being left, and if they were being honest with themselves, it hurt worse than leaving.

. . .

From the seat he had taken high behind technicians at TsUP—in the sort of balcony that old movie theaters had, with a grand view of the giant screens at the front of the room—O'Keefe had seen just enough to whistle out his first happy gust. He had been trying to keep his nerves in check when he had made small talk with Expedition Six and asked about how they wanted their steaks. But now, Bowersox, Budarin, and Pettit, having innocently come to represent the future of the American space program, were safe after undocking, and they were getting safer all the time.

O'Keefe's Russian counterpart, Yuri Koptev, was seated next to him. "Everything is going smoothly," he said, and O'Keefe nodded, a whisper more of the tension that had stiffened his spine easing out of him. He looked across at Bill Readdy and Paul Pastorek, once again by his side, and he smiled at each of them, but mostly O'Keefe smiled to himself. Expedition Seven's safe passage had been half of the gamble. Now, the other half—the half that had kept him awake for so many nights—looked as though it might pay off, too. Relief washed over him. Suddenly, he was exhausted.

He looked at his watch. It was nearly two o'clock in the morning. He had a little less than four hours before Expedition Six was scheduled to touch down on the Kazakh steppes. Watched pots and all that, O'Keefe decided to head back to his hotel and catch a short nap. He put his head on his pillow, and with the speed of a man who at last has nothing to worry about, he fell asleep, snoring softly.

. . .

Expedition Six made one and a half passes around earth in their tiny raft, waiting for the go-ahead to spark the de-orbit burn rockets that would slow their velocity only a little, but more than enough to drop them into earth's atmosphere. They filled the time taking salt tablets and drinking bags of water, trying to stave off the sickness that would follow their return to gravity. Pettit and Bowersox hoped aloud that their bladders would hold up. Following Russian custom, they had eschewed the diapers they would have normally worn.

In between swigs, Bowersox and Budarin, speaking to each

other mostly in Russian, tried to get a happy rhythm going in this new, old machine. They were having some trouble.

"We need to disconnect—" Bowersox began before Budarin put up his hand to interrupt him.

"Yeah, yeah, we will."

And then, while Bowersox read off various gauges, glancing every so often at his checklist to make sure the readings were what they should have been, Budarin—the expert—asked him about the order of things.

"Do we enter Format 45 now?" Budarin asked.

"Yes," Bowersox answered after consulting his book of instructions, as though they were about to build a piece of Ikea furniture, not about to rocket through space.

"Okay, we're entering that."

The seeming confusion prompted a different, harsher voice to cut in from Russian ground control. "Are you sure you know what you're doing? Are you all right?" The concern was justified. A delay of just a few seconds in firing the rockets could see Expedition Six land in the middle of the ocean rather than on the Kazakh steppes.

"Yes, we are sure. We are fine," Budarin said, and right on time, the rockets lit up, pressing the men back into their seats. They recounted for the technicians on the ground their decreasing speed, measured in meters per second: first they read out the length of time that had passed since they had ignited the rockets, and next they relayed their loss of velocity.

". . . 17 seconds, 18 meters per second," Bowersox said.

". . . 53 seconds, 23 meters," Budarin continued. "It's all according to schedule."

"Yes," Bowersox said, running his finger down a long list. "These numbers are exactly right."

". . . 1:05, 29 meters . . . 1:17, 35 meters . . . 1:25, 38 meters . . . 90 seconds . . . all parameters are normal . . . 2 minutes, 55 meters . . . fuel is okay . . . everything's okay."

Russian ground control urged him to keep reporting the time

and change in velocity. "Give the impulse," they said. "Don't worry about the fuel."

"... 3:15, 90 meters ... 3:20, 91 meters ... 3:30, 96 meters ... 3:35, 99 meters ... 3:40, 101 meters ... 3:47, 104 meters ... 3:50, 105 meters ... 3:55, 107 meters ... 4 minutes, 110 meters ... 4:05, 112 meters ... 4:10, 115 meters ..."

And with every passing count, *Soyuz TMA-1* inched closer to falling out of orbit.

Bowersox continued reading from his instruction manual. It was concerned only with the technical aspects of flight, not the emotional repercussions of it. It said nothing about what they should be feeling. "It's written here—" Bowersox said, lifting the page toward Budarin so that he could see.

"Yes, we will do that," Budarin said.

Bowersox continued running his finger along the pages. "On page 95, it says we need to wait until—"

"That is already open."

"Display off," Bowersox said, reaching out in front of him to flick another switch. "I'm turning to page 96."

"How is the pressure?" the ground asked, knowing that with the turn of only another page or two, Bowersox, Budarin, and Pettit would no longer be astronauts.

"The pressure is good," Budarin replied.

"The engine is working very smoothly," Bowersox announced above the last noise of the firing rockets. He sounded a little surprised.

· · ·

From their vantage point on the ground, the technicians at TsUP dug in for what looked like another routine return for *Soyuz*. Lines of expected data ran across their monitors. Everything was normal. Everything was clockwork. Everything was quiet, as it was in the dark city that slept around them, most of Moscow having turned in along with Sean O'Keefe. But spring was creeping up on them, and the nights were shorter now, and soon dawn would break. The day was forecast to start with clouds and rain, but even bad weather couldn't dampen the good feeling that summer was finally on its way. After all, the rain would only help with the melt. The back of another long winter had been broken. There would be fewer mice shivering in corners, huddling against the cold, for the cats to catch.

. . .

After the burn, the astronaut trio settled a little more deeply into their seats, having pushed through the first real force that had been applied to their bodies since their liftoff in the shuttle. It was exhilarating, but it was also a shock to their systems, like jumping into a cold ocean after having spent a long afternoon lying on the beach.

"Everything okay over there, Don?" Bowersox asked in English.

"Yeah," Pettit said, trying to find the right words to describe the sensation. "That was a nice kick in the pants, you know?"

"It feels like an afterburner lighting, doesn't it?" Bowersox said, recalling their times in jets, a feeling that for him, at least, was like a green flag, a signal that he was about to enjoy some action.

"Yes, that's a good description of it," Pettit said. "It feels like an afterburner . . ."

Pettit was suddenly distracted by the luggage sitting on his lap. He wanted to do something with it before things really started getting heavy. "So it looks like we have about five minutes," he said, referring to the countdown until *razdolina*—the forceful separation of the orbital and propulsion modules. Soon, there would only be their little bell.

"I have a whole bunch of stuff which I'll shove up underneath a cosmonaut panel—"

"What do you have?" Bowersox asked.

"I've got *neshtatny*," Pettit said, using the Russian word for a few of the books he held, detailing what to do in an emergency.

"Oh, you've got your *neshtatny* and Nikolai's *neshtatny*," Bowersox said, looking across at his weighed-down friend.

"Yeah, and I've got a reference book," Pettit said, holding up a thick *Soyuz* manual.

"Do you want to give them to me?" Bowersox asked.

"I'll find a place for them," Pettit said, a little proudly. He knew that he was the third member of a three-man crew, the closest thing to ballast among the breathing things on board. But he didn't want to be a burden, and he didn't want to be carried.

Just then, it started to look as though all three of them would have to be more independent than they might have thought. While Bowersox and Pettit were talking, Budarin was running into some problems with the radio. It was cutting in and out, and whenever it happened to be in, it was next to useless. Budarin and the ground filled the clear patches by asking each other again and again whether they were getting through.

"Can you hear us?"

"Yes, can you hear us?"

"Yes, can you . . ."

At last, the communication lines opened wide enough for Russian ground control to ask how the trip was going.

"Everything is good," Budarin replied. "Everything is calm."

In English, Bowersox and Pettit began thinking through the rest of their return: what happened when, what happened next. After the orbital and propulsion modules were jettisoned, what was left of their ship would rock back and forth all the while it dropped toward earth, quickly at first, but slowed by friction. A drogue parachute would open. A much larger parachute would open after sixteen pyrotechnic bolts exploded, and their capsule would float down more gingerly. The heat shield at the bottom of their capsule and the outer pane of their windows would strip away. Their seats would cock into a more upright position. A second before impact, just a meter from earth's hard surface, six small rockets would fire to soften their landing. And finally, a thump. The rescue teams waiting for them would spring into action, opening the capsule's hatch, lifting Bowersox, Budarin, and Pettit free of their cradles, and ferrying them onto a military helicopter. It would fly them to a plane idling on a runway at Astana, Kazakhstan, which would, in turn, fly them to Star City, where their wives would be waiting for them.

No matter how often they ran through the stages of their flight, the last one always involved hugs.

"Ken?" Budarin said, interrupting.

"Da?"

"We've done half our task," Budarin said in Russian. "Now we're really going to have some *fun.*"

Budarin spoke the last word in English. In a telling linguistic void, there was no Russian equivalent, and both Budarin's sentiment and his expression of it made Bowersox laugh.

"Make yourselves comfortable," Budarin said, smiling in return.

The ground cut in. "For six minutes, you'll have a communications break from us, but it doesn't mean you should be silent."

"You'll hear us moving around," Budarin said, "trying to squeeze into our seats. Our legs are cramping." Looking across at his crewmates, he said, "Make sure your visors are closed."

Until then, they could have kept their helmet visors open, hoping to stave off their feelings of being buried alive. The time for them to close up tight had come.

Bowersox, however, was occupied by other things. After he had switched off that display, it had occurred to him how easy it might have been for him to hit the wrong button—all of these small plastic squares lined up like bricks, tight against the next, and each of them looking, more or less, exactly like the others.

"You know, there's a good chance you might hit the wrong button," he said to Budarin.

"There's a chance," Budarin said. "I know what you mean. But don't worry. Everything's okay." He squinted at the instrument panel. "Check your visors, guys."

They did.

"Visors are closed," Budarin told the ground. "Separation program is on. Separation in fifty seconds . . . in thirty seconds . . . in five . . . four . . . three . . . two . . . one!"

For Pettit, the next second was the longest of the flight. If the explosive bolts that held the orbital and propulsion modules in place didn't work, Expedition Six were done for. Their heat shield would remain covered, useless, and the added drag would prevent them from dropping ass-first. Instead, they would helicopter into oblivion, their hull punctured, invaded by fire, and burned, like *Columbia*, from the inside out.

At last, there was the telltale sound of detonation, like machine-gun fire.

"Separation!" Budarin shouted. "We have separation! Everything works!"

The ground was silent to their drama. "Can you hear us?" Budarin asked. "Can you hear us?"

Bowersox, Budarin, and Pettit didn't know it at the time, but these were among the last words they would hear from the ground: "Keep an eye on your internal pressure, guys."

"Can you hear us?" Budarin asked.

There was no reply.

. . .

Micki Pettit had arrived in Moscow the day before with her twin boys. Annie Bowersox had also made the trip. Now, lying awake in bed, staring at their Star City cottage ceilings, they tried to sleep, but they were both kept awake by the thrill that morning would bring. They had made plans to wake up early and head to TsUP to watch from their front-row seats the return of their husbands. They couldn't wait. They felt like those giddy girls in Times Square who had welcomed the sailors home.

Only the slightest ill feeling clung to them. Nikolai Budarin's wife, Marina, had announced that she would not be there, because for a cosmonaut, it was considered the worst kind of luck for his wife to wait for him with open arms. If she did, it was almost certain that she would never close them around her husband again.

Micki and Annie told themselves it was just more Russian hocus-pocus, more silly superstition. So much had gone wrong already. They were due for a change of luck. Having made it through a heart-stopping beginning and an interminable middle, they were owed an uplifting end. They were owed their champagne moment, a tickertape finish.

In her bags, Micki had even packed a pillbox hat.

. . .

"Look at that fire," Bowersox said.

He and Pettit each saw the lost modules roll out past their windows and begin burning up. They were glad for not having been in them. But they didn't know there was still reason for concern. They didn't know that were everything in order, they wouldn't have been able to see what they saw. They didn't know that one of the small rockets assigned to keep their capsule stable had fired less than a second too late.

And then their own windows filled with plasma and fire.

"There's so much fire," Bowersox said, filled with wonder.

"Yes," Budarin replied, sounding distracted. He knew that some fire was normal, a product of the heat generated by reentry, the capsule trailing it like a meteor's tail. But even for Budarin, the fire seemed brighter than normal, more intense. It might have been his imagination, but the temperature inside *Soyuz* also seemed as though it was on the rise. Sweat started to run into his eyes. Blinking it back and turning his head to sneak a peek through one of the windows, he said, almost to himself, "Yes, we are on fire pretty good."

Bowersox and Pettit both marveled at the glow. But in his concern, Budarin had grown deaf to their awe. He was scanning the instruments and gauges, one by one, trying to find something, anything, that wasn't right. Suddenly, his eyes grew wide when one of his monitors flashed in front of him, and a telltale light—called, ominously, the BS light—blinked on.

Bowersox saw it, too. Uh-oh, he thought.

Holy fucking shit was more like it.

The computers had announced that whether Expedition Six liked it or not, *Soyuz* was about to be pushed into a steep, ballistic descent. Instead of the usual semi-gentle fall into gravity's embrace, they were primed to enter an accelerated, lung-crunching dive into

elementary physics. There was no longer time for grace. For whatever reason, the hardware wanted them home, as soon as possible.
It was as though the three men had been loaded into that shotgun
of theirs and fired straight into the earth.

In English, Bowersox gave Pettit the red alert. "Don, BS is lit up,
and we don't know why," he said. Resorting to his typical under-
statement, he added, "It's probably going to be a fairly aggressive
entry."

Budarin noticed that something was wrong with the capsule's
left side.

"I didn't touch anything," Bowersox said.

Pettit, unable to ignore the hint of anxiety that Bowersox had
failed to stifle during his self-defense, began to worry out loud.
"Why the BS?" he asked in Russian.

Perhaps because of the stress of the moment, Bowersox replied
to him in kind. "We don't know, Don," he said, before switching
over to English. "Tighten up your belts as much as you can."

The three men began tugging on their restraints, trying to find
safe places for all of the loose things that were about to turn into
projectiles.

"We'll make it, guys," Budarin said.

"Kolai, you're good," Bowersox replied.

"Guys," Budarin said, trying to stay focused on the instruments
in front of him through a growing shake. "Hold on, guys, hold on."

. . .

Sean O'Keefe's alarm went off. He pulled himself out of bed and
tried to shake out the cobwebs. He smoothed down his hair, pulled
on a jacket, and headed back to TsUP.

Paul Pastorek joined him again in the gallery, as did Bill Readdy.
Micki Pettit and Annie Bowersox had also arrived, looking excited
and put together, what with Micki wearing her snappy hat. The two
women took their seats near O'Keefe, and he turned to smile at
them. Returning the smile, Micki and Annie leaned forward to get
a better look down at the floor of technicians below.

Russian ground control was staid and beautiful, all marble

columns and heavy drapes. Everything looked calm, as peaceful as a library. And on those big screens at the front of the room, a series of almost cartoonish illustrations was being projected, explaining what was happening to their husbands and when. According to the cartoons, everything was going to plan. *Soyuz* had dropped into the atmosphere and made a smooth, on-target descent. Now its parachute was about to open, and after the capsule had bounced to a happy stop on a forgiving earth, what looked like Yogi Bear and Huckleberry Hound were ready to jump out of it. Perhaps they would have a picnic.

. . .

"Can you hear us?" Budarin repeated again and again. But still there was no response from the ground. Expedition Six were alone.

Budarin's breathing grew harder. "Tighten up as much as you can," he said through gritted teeth.

Bowersox licked his lips. Pettit closed his eyes.

Budarin had leveled his sights to a single gauge in front of him, the needle in it bouncing and rising slowly, recording the g-forces that had begun to sit on their chests like barbells.

"We're at 2.0," Budarin said, a little nervously, "2.3 . . . Hold on guys."

"We're holding on," Bowersox said in Russian. And then, in English, he said to Don: "Take a deep breath while you can."

The capsule had begun to spin. There was noise, snaps and rattles and groans, and vibration, each rising in pitch. Outside, the fire and plasma danced, coating their windows with ash. Alarm bells went off, if only in their minds.

"Don, how are you?" Budarin asked. "Speak so we can hear you."

"Da," Don said.

Budarin continued the count. ". . . 3.0 . . . 3.5 . . . 3.9 . . . Don, speak to me, say something to me."

"Da," Don said again, this time croaking it out.

Their spines compressed. Their ears rang. Pettit could feel sweat streaming back from his forehead and soaking his hair, as though he were in a centrifuge. Bowersox fought to keep his tongue from slipping down his throat.

"... 4.0 ... 4.35 ... 4.44 ... 4.7 ... oh, it's pressing good."

Already, nearly a thousand pounds sat on each of their chests, and things were only getting worse. With every second it grew harder and harder for them to breathe, their gasps already short and shallow. It took everything in Budarin for him to continue to talk.

"... 5.0 ... 6.0 ..."

They approached the g-force limits that the human body, if left in a vulnerable position, can survive for any length of time. After nearly six months in space, weightless and free, for this brave trio it felt like torture, as though some maniac wanted to see how far he could push them before they finally broke in half. Budarin continued to talk, but soon his audience had trouble listening. So much blood had been pushed to the backs of their brains that Bowersox and Pettit felt as though they had been sucker punched. Were they not already flat on their backs, they would have been knocked there.

"... 7.0 ... 7.5 ... 7.9 ... 8.0 ..."

Now Expedition Six had reached an almost mythical number. Several racetracks have been redesigned because drivers in their new, faster cars have reached 5.0 in the corners and risked passing out and crashing. At 8.0, Bowersox, Budarin, and Pettit were sustaining an occasionally lethal level of crush, one that threatened to pinch their weakened lungs shut tight. They couldn't have been blamed if they had panicked. This was one more surprise that they could have done without.

Fortunately, inevitably, *Soyuz* continued its fall through the atmosphere. Warmer, denser air began slowing them down, and right when they needed it to, the weight began to lift.

"... 7.6 ... 7.5 ... 7.1 ... 6.5 ... It's great," Budarin wheezed, "... 4.3 ... 3.5 ... 3.11 ... 2.8 ... 2.2 ... 1.7 ..."

Bowersox and Pettit blinked back their fogs. Their blood began rising back into their faces, their tongues meeting their unclenching teeth. They took great, gulping breaths, as though a bully had just taken his foot off their necks. Most important, they even found it in them to smile, having passed one more test, but with more to come.

"Don, get ready for the parachute," Bowersox said.

"Okay," Pettit said, weakly.

"It's easy now," Budarin said. "Now we'll have *fun* again."

Bowersox, however, wasn't yet thinking about spreading out a blanket in the sunshine. Instead, he was busy pouring all of his might into willing the parachute to open. By the book, it was part of the *Soyuz*'s automated operation, and he bristled at the lack of control—not just the pilot in him but the hardened realist in him who had survived one malfunction and didn't fancy his chances of surviving another.

Just then, the small drogue chute opened, filling with air. But the huge main chute didn't follow its lead. The pyrotechnic bolts that kept it folded tight still hadn't fired.

Bowersox shook his head. He wished for a huge red button to appear in front of him that he could press, hard, and more than once, to release the parachute. But there wasn't one. There was just the cruel wait, while Expedition Six continued their race toward the cold, hard earth. The gauges showed the capsule was traveling more slowly than it had been, but when it comes to falling out of the sky, pace is a relative thing. The three men were still going plenty fast enough to dig their own graves.

. . .

In Moscow, where officials anticipated *Soyuz TMA-1* to make its gentle touchdown within sixteen minutes—how close to home *Columbia* had been when it was lost for good—the radios came back to life just in time to broadcast a short, loud blast of static. Then the radios crackled, and then they went dead.

In the silence, a few of the technicians put their faces into their hands. A few of the others looked snow white.

Because sometimes, bad things can happen twice.

The Americans in the gallery weren't all that alarmed by the stern masks suddenly put on by their Russian colleagues. Nor were they unsettled by the tense quiet or by the occasional arrival of a harried-looking subordinate, whispering into the ears of one superior or another.

The cartoons, after all, were still showing happy scenes of a flawless flight, and that was their singular focus; everything else was a mystery. Locked away in this great room in the dark, and unable to speak Russian or make out the whispers, they had been dunked into a kind of isolation tank. Unlike that morning when he had waited hopelessly for *Columbia*, Sean O'Keefe couldn't see a touchdown clock counting past zero, couldn't worry about the sonic booms that he hadn't yet heard, and couldn't read the fear in anxious faces. All he had to go on was what he could see, and just then, just there, he could see only cartoons and a kind of stage play, a crew of silent actors running through their routines. It was as though he was stuck in the back of a darkened theater, watching an opera in a language that he couldn't quite follow. And so he sat, along with the others, blindly waiting for the aria.

They were oblivious to the possibility that three of the principal parts were being played by fire and smoke and ash.

Finally, O'Keefe, Pastorek, and Readdy—as well as Micki and Annie—saw the screens at the front of the room fill with grainy color footage of a *Soyuz* capsule thumping into the steppes, kicking up dirt. Its orange-and-white parachute rolled out in front of it in a gentle breeze, flapping like a deflated hot-air balloon, and within minutes, soldiers and technicians huddled in helicopters had spotted it and touched down nearby. The film, in essence, showed a textbook landing and recovery unfolding. In the balcony, there was relief. All that remained was the cracking of the hatch.

But suddenly an open radio transmission that had been playing for the assembled crowd, which now included a number of Russian reporters, crackled with the concerned voices of search pilots who hadn't yet caught sight of Expedition Six's parachute. The Americans were confused by the seeming discrepancy between what they were seeing in front of them and what they were hearing through the radio. How could the helicopters be touching down if the pilots weren't sure where *Soyuz* had landed? And who was taking these pictures?

And then it dawned on the group of them—not quite all at once, but instantaneously enough for a feeling of dread to spread like a virus through the balcony gang—that the footage that they had been watching was stock. It was one more of those cartoons, just without the choppy animation, real life turned into make-believe.

Just then, there was a buzz among the Russians, some of whom had begun to sweat. When they weren't listening to the radio, they spoke mostly in hushes. They weren't speaking in hushes anymore. Finally, after ten minutes, Yuri Koptev (trying his best to summon a sense of calm) told O'Keefe that the pilots hadn't seen the parachute because *Soyuz* had overshot its landing site by a few kilometers. No doubt it would be spotted presently.

After another fifteen minutes had passed, O'Keefe was told that *Soyuz* had fallen short of its target, perhaps by as much as sixty kilometers. For the Americans, the uncertainty was bewildering at first and made them feel sick second.

When everything goes according to plan, *Soyuz* lands in an area precise enough for the rescue teams to watch its parachute open.

This time around, there had been no sighting. There was only an empty sky.

. . .

After what felt like an eternity, Bowersox, Budarin, and Pettit heard what they thought were blessed pops—the sweet sound of the bolts exploding. The noise echoed through *Soyuz*, the capsule shuddering.

"I think the parachute is opening!" Budarin said, following his exclamation with a whoop, like a cowboy at full gallop. "It's time to hang on again, guys! Hang on!"

. . .

Finally the Russians told O'Keefe that, all apologies, they would have to excuse themselves for a moment. Koptev hoped to leave a feeling of reassurance in his wake, but he had failed. There had been too much rush in his strides. The Americans, feeling helpless and being watched too closely by the reporters who surrounded them, also decided to withdraw. They set up camp in two otherwise unoccupied rooms, one of which was a small, drab kitchen. With worry filling in the gaps between the huddled bunches, the two rooms felt like bomb shelters.

Someone put on some coffee. Someone else paced. More ominously, Pastorek pulled up a chair to a table and began taking notes. For O'Keefe, watching his friend's pen dash across the page was like being a ship's captain watching a fast-moving blip on his radar screen closing in. Nothing good had ever come out of either. Each was notice of an incoming torpedo.

In his growing upset, O'Keefe was immediately returned to the side of that empty runway in Florida. He flashed back to that morning—to seeing the stricken faces of seven families before they were hustled out of view, to knowing that his prayers for them would go unanswered. He remembered the awful days that had followed, the mourning and debate and recrimination. He remembered the memorial service. He remembered the flowers and cards and teddy bears that had been piled against the front gates in Houston. And now he saw them as though they were right in front of him. He saw every last bit of it, and he saw the beginnings of it happening all over again.

Every now and then, he sent an emissary over to the Russian side for an update. Each time, the emissary came back with a look on his face or a shake of his head that brought tears to Micki Pettit's eyes. She had decided that she needed to beat a further retreat and ended up hiding out in the relatively quiet kitchen. Even the unflappable Annie Bowersox, who repeated again and again that everything would turn out all right, that everything had always turned out all right, sounded a little less sure of herself with each passing minute. Everybody in each of those two rooms knew that the longer the mystery lasted, the more likely it would end in a cemetery.

O'Keefe waited, the big man shuttling between those two semi-silent, semi-hysterical rooms, until he couldn't take the pounding in his ears any longer. He needed some time alone. He also needed a cigarette. He slipped through one of the doors and disappeared.

Pastorek looked up from his notes and saw that his boss and best friend was missing. He got up from his seat and left the room, looking down long, empty corridors and poking his head through open office doors. He spied O'Keefe at the end of a hallway, staring through a pair of corner windows. Smoke curled from his fingertips.

"You okay?" Pastorek asked after he'd made a tiptoe approach. He knew full well the answer.

O'Keefe shook his head and took a drag from his cigarette. "I just can't believe it," he said finally, blowing out a thick stream. "It's like *Columbia* all over again. It's déjà vu all over again. And the worst part is, we don't even have our own people in control. I mean, look at this," he said, waving one of his big hands at the vista through the windows.

Rising in front of them was the rusted hulk of a building that was either unfinished or abandoned (or probably both), all un-capped girders and cracked foundation. Grass grew through the crumbling sidewalk that ran past it. Dark clouds and the forecasted rain had started blowing in. Everything else was already a shade of gray, one long shadow having crossed the city. It was the sort of landscape that looked as though it had never once been touched by rainbows.

It was such a foreboding view—so close to a greasy postapoca-lyptic movie set or the pages of a dark graphic novel—that Pastorek almost cracked a smile. "It's an unreal place," he said. "It's like be-ing in Wonderland."

Both men stared through the window for a few more beats be-fore Pastorek turned back toward O'Keefe. "Where's your thinking at?" he asked.

O'Keefe took another long pull from his cigarette.

"I think we're in trouble," he said.

. . .

The capsule bounced and shook while its massive orange-and-white canopy blacked out the sky above it. Now, Expedition Six's long fall was finally on the verge of being broken. Despite having enjoyed so long a stint of freedom, the three of them had never been so glad to be tied, once again, to the end of a string.

"It looks like we're going to live for another day," Pettit joked in Russian, now that the capsule's fall was slow and smooth.

"Yes," Budarin said. "Life is getting better."

"Everything is *nishtyak*," Bowersox said loudly, throwing down some Russian slang he'd picked up along the way, something like *cool*.

"That's right," Budarin said, smiling. "Congratulations, guys. We got through 8.0. It will be easier now." And then, in English, he said, "Medical test is done."

Now, with fifteen minutes remaining until the capsule touched down, Bowersox, like every good test pilot working through the kinks that he'd found in his new machine, wanted to begin the long process of determining what had gone wrong and why. He didn't feel as though he had made a mistake—as though he had pushed the wrong button or accidentally toggled a no-no switch—but some small part of him wondered whether he might have been responsible for the rough ride. "But how? I guess we'll find out," he said, before an unfamiliar voice interrupted his thoughts, squawking across their radio.

"I am the search plane," the mystery voice announced. "Can you hear me?"

"Search plane, we hear you," Budarin replied.

Silence.

"Search plane, come in, reply," Budarin said. "Answer us."

More silence.

Until, finally: "We're glad to welcome you. Search teams are in the area. Did you receive a weather report?"

"Yes, we received it," Budarin said. Unlike in Moscow, it was going to be cool on the steppes but sunny, blue skies dusted with a few white clouds. But more pressing in Budarin's mind was their altitude, from which they could calculate when they might touch down. Normally the search planes were in sight of *Soyuz* and could pass along an estimate. "Can you tell us our altitude, so we can prepare ourselves for landing?"

"When we see you, we'll let you know," the pilot replied. "For now, all we are getting are radio signals."

"Okay," Budarin said, although he wondered how the search planes, rocketing over the flats, couldn't see their smoking capsule through clear, bright skies, floating under its candy-colored parachute.

He didn't dwell on the dilemma for very long. Instead, he grabbed Bowersox and Pettit by their knees and shook them. "I congratulate you, guys. We got through it. There is just one more moment to survive."

Budarin was talking about the landing.

• • •

The clouds grew darker in Moscow, the rain heavier. Rumors and conflicting reports began filtering through the walls, carried by the vents, back and forth between the Russians and the Americans, the boardroom and the kitchen. There were several NASA representatives in the search helicopters and waiting at the target site—including the normally reliable Bill Gerstenmaier—but the jet engines and rotors washed out phone transmissions, and the few swatches that made it through were ratty and incomplete. Frustrated, some of the Americans at TsUP began calling Mission Control in Houston for news. Perhaps they were hearing things that weren't making it through the low ceilings in Russia. But they were not. There was nothing to report. No beacon, no voices, no sighting.

Expedition Six had vanished.

• • •

"What's our altitude?" Budarin asked again and again.

"We can't see you," the pilot said each time. "We'll let you know when we see you, but we can't see you right now."

That's impossible, Budarin thought. The planes should have easily spotted them by now. The ground was rushing up to meet them. In the past, there were always rescue teams waiting, ready to catch him. Where were they now?

Bowersox, still serving as go-between, began preparing Pettit for impact, in English. "A few seconds before landing, Don, you'll see the landing light go on. That means, you know, get yourself in position. You know that."

Just then, the capsule's vents opened up to equalize the pressure between outside and in. Vapor filled the air like so much cigarette smoke, and condensation dripped from the gauges and instruments. Their ears popped. They were seconds away from their final moment of truth.

"We need to get ourselves ready," Budarin said. "Hold on to your seats." Then, turning his attention back to the missing search planes, he asked them again to check their altitude.

"I hear you, but I still can't see you," the pilot answered. "We think you've missed the target by forty kilometers. We'll be there soon, but right now, we still can't see you."

"Everything is different for us," Budarin said, shrugging toward his American friends. Because the search pilot seemed sure of Expedition Six's location, no one on board thought to mention the ballistic descent.

"I can hear you, but I still can't see you," the pilot repeated.

"Okay," Budarin replied. "We are in a good mood, we're fine. We're getting ready to land."

He asked Bowersox to try to wipe off a gauge that had fogged over. It calculated the pressure of the outside air, and it had been counting up while they had dropped out of the sky. Once they were on the ground, it would read 731. Until then, it would help them take a stab at their altitude (although why there was no altimeter inside *Soyuz* was anyone's guess). The closer they came to 731, the closer they were to home.

"It's at 525 now," Bowersox said, knowing that it wouldn't be long now. "Don, are you comfortable in your seat?" he asked in English.

"Da," Pettit said. "Da."

"The pressure is at 660," Bowersox said. He could feel even that small difference in his legs and gut, the weight of it pressing him down into his seat. So, too, could Pettit and Budarin—a gentle reminder of those harrowing moments when each of them had felt the weight of the world on his chest.

"Well, it's nothing compared to 8.0," Budarin said.

"I only remember seeing 6.0," Bowersox said. "I think I lost track after that."

"Yes, I think so," Budarin said. "After six months 8.0? That's pretty good."

"It was good," Bowersox said distractedly, his mind having already begun tackling another problem. If the search planes couldn't see them and if their descent had been that much steeper than it was meant to have been, then perhaps they had fallen well short of their target. "Are we still in Kazakhstan?" he asked Budarin, only half joking. "Is this going to get political? Can we get to Ukraine this way?"

"No, no," Budarin said. "We're where we're supposed to be. We are still in Kazakhstan." He had kept his faith in the search pilot's announcement. "Either way, we're landing soon. Get ready."

Bowersox relayed the message to Pettit. "Get ready for landing anytime, Don." Bowersox did some quick math. "I think we're still around 10,000 feet, but—"

"Roger that," Pettit said.

The search plane pilot, still unable to catch sight of the capsule, tried to narrow his sights. It was now his turn to ask, "Can you give us your approximate altitude?"

"Our pressure is 650 now," Budarin said. "What's that give us? About 1,000 meters?"

There was no reply from the pilot. They had lost their radio connection yet again—*Soyuz* having dropped below the horizon—and now, there would be no getting it back. For the rest of their ride, at least, and perhaps for much longer, the three men of Expedition Six would have only one another for company, as though they had been launched back into space, not returned to earth.

Budarin asked Bowersox and Pettit if they could see anything out of their windows. They lifted their heads and had a look, but they couldn't see much. There were just flashes of sky and white cloud. Had they been flying their machine, they would have been flying blind.

"Pressure is 680 . . ." Bowersox said.

He continued to watch the gauge.

". . . 700 . . ." he said.

The reading continued to climb.

"... 710 ... 720 ... 730. It's soon—"

When suddenly, the capsule shuddered, and the men were thrown hard into their seats, their necks snapping back, their ears filling with the sharp clang of metal in distress and their noses with dust.

Bowersox spoke for Pettit, too, when he looked at Budarin with wide eyes and asked: "What the heck was that?"

11 THE WEIGHT OF THE WORLD

"That was landing," Nikolai Budarin said with a sheepish grin.

But it wasn't over yet. Before he could press the button that would cut loose their giant parachute—a button he would not have wanted to press even half a second too soon—a strong wind filled the canopy, and the scorched capsule began bouncing across the Kazakh flats like a tumbleweed. After a few seconds, Budarin managed to release *Soyuz* from its sail, but not before it had been tipped over onto its side. By the time they had come to a real, permanent state of rest, Ken Bowersox was on the bottom of the pile, Budarin was wedged in the middle, and Don Pettit was perched on top, feeling impossibly heavy, with his arms and legs hanging limp in the air.

"We're good, we're good," Bowersox said, after they had come to a stop and the dust had begun to settle.

"Guys, try not to move your heads," Budarin warned from previous experience. After spending so long in space—like sailors who have been out to sea one too many times—now they risked landsickness, risked having their insides squeezed out by gravity's crush. The taller Pettit, especially, was already feeling a little like Atlas. He didn't need to be told to stay still. Even blinking felt like too much hard work.

Bowersox, though, couldn't resist turning his head to look through the window just to his left. It was pressed against the earth, and through it he could see a few leaves of grass pressed flat.

"It's so good to see the grass!" he said. "It's so nice to see dirt!"

He laughed out loud, surprising himself at his joy, the mundane having turned for him into sculpture. The good feeling reminded him of the first time he had survived jumping out of a plane with a parachute strapped to his back. In that moment, after his high-velocity plummet had been slowed by his canopy and finally stopped by a forgiving earth, he had come to understand why skydivers kiss the ground. There was such happiness in having survived the fall.

"Let's go outside," Bowersox said, eager to get even closer to that green grass and brown dirt, to feel it under his feet, and to take in deep, cleansing breaths. But Budarin, believing that the recovery team was only moments away and not wanting to look to his comrades at TsUP like a maverick, parried the idea of cracking open the hatch.

Instead, following procedure, he pressed a button that prevented the capsule's several small radio antennas from deploying; each was locked behind an explosive hatch, and Budarin didn't want to pepper the search party, which he assumed was nearby, with shrapnel. What he couldn't have known, however, was that because *Soyuz* had tipped over onto its side, their main antenna had plunged straight into the ground. That misfortune left them more removed from the outside world than they had ever been up there. Even the pilots who had been talking to them minutes before could no longer hope to make contact.

The three men of Expedition Six were ignorant to the long-range panic that filled in the gaps in the static; they had no idea that every tracking device available had lost its hold on them, and that everybody on the other end of seemingly every radar and radio in Russia—pilots, engineers, wives—had lost their hold on them, too.

"Let's just sit here," Budarin said. "The helicopters should come soon. Let's wait until they rescue us."

To fill the wait—and, more important, because his checklist told him that he should—Bowersox recovered from his early excitement and began looking at the capsule's lights and gauges and writing down their final readings. He recorded the facts and figures of their incredible flight, but now more true to form, he was careful to keep the emotion of it to himself. Besides, there was no one to share it with. There were no friendly voices coming across their radio. There still hadn't been a knock on their door.

"Can anybody hear us?" Budarin shouted out into the emptiness. "If anybody's out there, we're on the ground. We're good.
We're waiting for your commands. We're waiting for rescue."

He looked up at Pettit and asked him how he was feeling.

"I'm fine," Pettit said, lying just a little. He was swallowing
hard through the world's worst case of bed spins.

And then they went quiet. It felt as though they had run out of
things to talk about, like three men sitting at a bar—checking their
watches, counting down to closing time—but with nowhere else
better to go.

"Usually, you have to wait about thirty minutes before they
come," Budarin said, trying his best to make the delay seem normal.

But Bowersox, his mind having never stopped turning over, was
only half listening. Already he had decided that the rescue crews
were still hours and maybe even days away. He had done the ballistic math and calculated that he hadn't been far off when he had
joked about landing in Ukraine.

Still, he told himself that, in some ways, lines drawn on maps—
and which of them they had fallen within—were irrelevant. No matter whose grass was pressed flat against his window, Bowersox
remained strapped inside a small pocket of Mother Russia, an hon-
orary citizen and a guest of a cosmonaut. *Soyuz* was Budarin's turf,
and Bowersox was mindful of the necessary diplomacy for as long
as he was in it.

And yet he didn't especially want to lie twisted and cramped until their rescuers finally showed up.

"You know, it could be that we're four hundred kilometers
away from our target," he suggested with all of the gentle humility
he could muster.

"Yes, the search planes couldn't see us," Budarin said, thinking
back to their last, confused communications with the outside world.
But just as it seemed he was about to reach out for the handle of the
hatch, he pulled back his hand. "We don't want to move around.
Let's just rest."

They remained suspended in silence, punctuated only by a sporadic debate between Bowersox and Budarin over whether to switch off some vents that had been running since they had first depressurized the capsule. Bowersox's manual dictated that they should, but Budarin wanted to keep them on until they headed outside.

"It's written here—" Bowersox said again and again, pointing to a page in his book.

But Budarin refused to look. It was his turn to be the commander, and, as he had in his decision about the hatch, he wanted it known that his word was final.

Once again, Expedition Six fell into an almost uncomfortable silence. After nearly six months of tranquillity, they had found their first grounds for conflict. It had taken gravity to remind them of their differences.

"The main thing is, we didn't hit the water," Budarin said. "At least we hit the ground."

"Absolutely," Bowersox said, seizing the opening for dialogue. "And we're lucky that it's cool outside. But I would prefer to leave. What if we waited for an hour? Then can we leave?"

Budarin sighed like a parent who was being hounded by a pestering child. "We have to calm down a little bit," he said. "We need to compose ourselves, and then we'll see."

. . .

Back in the bomb shelters, Sean O'Keefe wondered aloud, and a little angrily, why there wasn't a fucking satellite phone in the fucking *Soyuz*, so that Bowersox, Budarin, and Pettit could call in and announce their fucking location. (Russia's safety engineers, it turned out, were unhappy with the batteries that powered such phones; they worried that they might leak toxic gases for as long as they were stored in space.)

Paul Pastorek thought about trying to answer the question, but instead he continued taking his notes, having decided that his boss didn't need to hear that a satellite phone would have been of use only if Expedition Six were alive to use it. His growing pessimism matched the collective mood in those two rooms. Micki Pettit—

jet-lagged, tired from raising two sick children on her own, and stressed out of her scrambled mind—felt as though she might spontaneously combust. It had already been too much for her to take. Now in the waiting, she was pushed closer to the ends of her earth.

. . .

"How many g's do you think we pulled there?" Pettit asked in Russian, the first words out of his mouth since he had lied about being fine.

"Nikolai said 8.0," Bowersox answered in English, almost wistfully. "I hope I didn't make a mistake," he said, switching back to Russian.

"It's fine, guys," Budarin said. "We're on the ground. That's all that matters."

"I didn't touch anything," Bowersox continued, not hearing. "And you didn't touch anything. It should be automatic. I'm worrying—"

"It's okay, Ken," Budarin said again. "Don, how are you?"

"So-so," Pettit said. "A little—"

"It's okay," Budarin said, sounding again like a parent, although this time a gentler one. "You'll be home soon."

His own yearning to be there made it harder for Budarin to remain patient. He picked up the radio again. "We're on the ground, we're safe, but we don't know where we are," he said. "Does anybody hear us?" After listening to a wash of static, he turned to Bowersox and asked if he could see anything new.

"No, just the grass," Bowersox said, turning his face toward the window. "But it's so beautiful."

In orbit, colors had been muted by filters—by distance, by space, by the atmosphere, by clouds, by dust, by having spent so long away that red and yellow and green had been reduced to memories. Now here was color again, and that grass was so bright, so green against the deep brown of the earth, that Bowersox couldn't help staring at it. He saw home in those leaves. Each time he looked at them, he felt more and more human, the robot come to life.

Pettit, though, was having a harder time adjusting. He felt hot and heavy and tired and sick. "Sox," he said in English, "I'm not going to be much good outside."

"No problem, buddy," Bowersox said, still staring at the grass and sounding enraptured. "We'll just get you in the *palatka*"—the tent that cosmonauts are traditionally carried to after they're lifted from *Soyuz*, to protect them from the rain and the wind that they haven't felt for so long—"change your clothes, and let you go to sleep on the airplane." It all sounded very pleasant.

Pettit wasn't convinced. "I'm afraid I'm going to lose the package," he said.

With that revelation, Bowersox suddenly felt less sunny. Gravity was set to bite him in the ass a second time. Still stuck at the bottom of a three-man pile, he stared down the prospect of Pettit showering him with the dregs of his last space-station chow.

"That's okay," Bowersox said, although the way he said it made it clear that it was not entirely okay. "Just do it in the bag."

Pettit was silent for a beat, as though trying to resist the urge to puke. Instead, he was trying to figure out what Bowersox meant. When Pettit had said that he was going to lose the package, he wasn't trying to find a gentle way to say that he was about to barf. He was literally losing hold of the books and manuals that had been pressed to his lap, now squeezed between his slippery hands. "No," he said, "I'm talking about the—"

"Oh!" Bowersox said, laughing with relief. "That's okay. Give 'em to me. It's my turn to hold 'em for a while." Pettit passed the bundles down, and Bowersox tucked the books between his head and the capsule's inside wall. "They make a great pillow," Bowersox said.

With perhaps the worst crisis of the entire mission averted—the cabin, for now, remaining vomit free—Budarin turned his attention once again to raising the search planes, still to no avail. "How long have we been waiting?" he asked.

"About twenty minutes," Bowersox answered. "Do we have a beacon, a light we can shine?"

Budarin shrugged, lost in thought.

"Are we going to open the hatch?" Pettit asked.

Budarin didn't answer, but after taking a long look at the Russian, Bowersox did: "No, not yet, Don. But Nikolai's thinking about it."

The sun was back in his voice.

· · ·

Bill Readdy had seen the tension ratcheted up high enough. In his cool, quiet way (he had a knack for making his usual monotone sound even more monotone when needed) he took it upon himself to try to calm down his friends and colleagues. He knew that because of *Columbia*, everybody was already on edge, already chewed down to the nubs. But that was then, and this was now. Calling the American delegation to his attention, he explained that it was perfectly normal for the radio transmissions to cut out. It happened all the time. Perhaps the capsule was on the other side of a hill, or perhaps its antenna had broken off. And had Expedition Six entered something called ballistic descent—he didn't get into it, but it was his gut feeling that they probably had—they were no doubt fine, but they were sitting on a patch of grass a long way from their target. It would take longer than anyone might have liked for the good word to come through, but that good word would come, not to worry.

His speech bought a few minutes of much-needed downtime. There was quiet. But soon enough, phones began ringing again, and people began pacing and shouting again, and for most of the men and women in those two rooms, it started to feel as though the walls were closing in.

· · ·

After more than forty minutes locked away, restless and stiff, Bowersox tried one last time to get Budarin to change his mind. "I think we should get out," he said.

"Yes," Budarin said. "I think we should get out."

At last, he reached across and cracked open the hatch. Sunlight and fresh air streamed into the capsule. The light was so bright that it almost hurt, but the air was clean and beautiful. They drank it up like spring water. In space, they had taken in a higher concentration of carbon dioxide than they were used to, and for Bowersox, it had left him feeling a little less like himself—as though he had a toothache or hadn't had enough sleep. Now, swallowing down great big gulps, he was reminded of how he felt when he was in his space-suit, outside station, filling his lungs with pure oxygen. He had felt alive then, just as he felt alive now, happy and content. He drew in a calm that he hadn't known for a long time but had learned not to miss.

One by one, they unbuckled their seat belts, dragged themselves toward the hatch, shimmied through, and fell to the ground. There were no wolves. There were only white birds against a blue sky and Kazakhstan's steppes, stretching out for nearly three hundred empty miles between them and their target. In that gorgeous moment, it could have been just a few feet or even a few million miles. Either way, it didn't matter to them. There was a good possibility that in the long history of the planet, no one had ever been where they were. But for Expedition Six, this baleful place felt as familiar as breath. The sky, the grass, and the birds, they were universal. They were the constants that they had missed, the reminders that each of us see every day but rarely stop to take in, the signposts and flags and whispers that tell us that no matter where we are, on some level we are home.

"It's really, really beautiful," Bowersox said.

Budarin and Pettit nodded. Other than those nods, the three men stayed perfectly still for more than an hour, flat on their backs, feeling the sun on their faces, watching the birds. They were like kids staring at clouds. It was hard for them to know it then, but when they would come to look back on that time, they would re-member it as one of the most perfect hours of their lives.

But with no call on the radio and with no shadows on the ground, they decided that they needed to start thinking about trying to rescue themselves. Bowersox and Budarin took a stab at stand-ing, but the effort only made them feel sick. Instead, Bowersox pulled his spacesuit halfway off and crawled back into the capsule to retrieve supplies, first pulling a fleece sweater from the survival gear behind his seat. He also passed Budarin the shotgun, some flares, and a beacon that he'd found. It was set up, and then Bowersox began calling out through the radio, straining to hear an echo in the static.

"Knock knock. Anybody home?" Bowersox said. "We're still waiting. Please, guys, find us."

He called out again and again until, two hours after *Soyuz TMA-1* had landed, one of the search plane pilots came through. "Can you hear us?" he said.

"We hear you," Bowersox said, slumping a little with relief. "We hear you okay. Do you hear us?"

"We hear you well," the pilot said. "How do you feel?"

"We're tired," Bowersox said, the adrenaline hangover having started coming on strong. "I feel like going to sleep. But it wasn't a bad trip for the first time."

Just about then, Bowersox realized he had been talking to himself. They had lost the connection.

After a long while, the radio crackled back to life. "How do you feel?" the pilot asked again, having missed the answer the first time around. "Sorry," he was quick to add, not waiting for Bowersox's reply, "but we've had to touch down. We're out of gas. We'll talk to central command and leave again soon."

A little disgusted, Bowersox put down the radio. Despite the breakthrough, the pilots still had no idea where to find Expedition Six. They could not be guided over that sort of distance by voices. And now they weren't even in the air. But sore from lying across the lip of the hatch—and because he is the sort of man who would try to walk off polio—Bowersox finally mustered the strength to stand up, stretch his back, and take a few tentative steps.

Nearby, Pettit stayed put, only every so often attempting even to sit up, leaning on his elbows, but he kept busy all the same, thinking through a way to turn their parachute into a nifty shelter for the night.

Budarin, meanwhile, heard what he thought were cars speeding in the distance. He, too, forced himself to stand, loaded the shotgun, and began trotting in the direction of the noise, firing flares into the sky. They rained down in thick red clouds, but it turned out that Budarin had been hearing only gravity's tides rolling in, and nobody came.

All the while, their beacon had begun sending its signal deep into space, past station, toward a satellite orbiting thousands of miles above the earth. And now that satellite had sent back the signal, like a father playing catch with his son.

. . .

It was announced at TsUP that the faint sound of a beacon had been heard. Search planes were scrambling in its direction. But for both sides, that news was not met with any sort of enthusiasm. Beacons were one thing. Human beings were another. Plenty of times, parts of machines had survived when their crews had not. The Russians remembered that when the three members of the Salyut 1 crew had been lost, their capsule had been intact, and their beacon had sung out loud and clear. And even if the news of the beacon were true (the Americans weren't so sure that it was), they knew that faint beacons had called out from places where men never could—sunk to the bottoms of oceans or trapped under ice. There was no way to know whether that high-pitched squeal was announcing an astronaut's safe arrival to life on earth or his premature departure from it.

There was no way of knowing, either, what happiness or heartbreak the next phone call or announcement might bring. Though it remained largely unspoken, and for all of Readdy's kind counsel, there was a feeling among some in those rooms—between O'Keefe and Pastorek, at least—that they would have to start talking about what would happen if the coin flipped heartbreak.

Casting nervous sideways glances at Micki and Annie, tearful and huddled together, O'Keefe wondered how they and NASA, and perhaps even America, could ever make it through such a loss. He began formulating the steps that he would take to try to save the women and the souls of their husbands and the dreams of Mars. But each time O'Keefe tried to find his way along the path to redemption, his sleeve got caught on a bramble or he lost his footing, and he had to start all over again. He never did reach the end. Try as he might, he couldn't bring himself to run through the logical course: shock, then sadness, then anger, and then a lifetime of grief.

Instead, he went out to sneak another cigarette, and to chew on his mustache, and to run his hands through his hair, and to pinch the bridge of his nose, and to rub his eyes. When it came right down to it, these were the only things he could do. It was all he had in his power. This was all he had left in him.

. . .

Just when they began thinking that they would have to spend the night, cold on the flats, Expedition Six heard the screaming of jet engines. Brought to them by the signal, the first of fifteen search planes appeared high overhead, nothing more than a jagged shape against the still-bright sky, dipping its wings when it finally caught sight of them, three men laid out on the grass. The pilot radioed down.

"Sorry for the wait," he said. "The helicopters are coming. The main helicopter is two hours away."

"No rush," Bowersox said back, and he meant it. For the first time in months, maybe even years, he could genuinely relax, his work finally done. "Say hello to our wives," he said as dreamily as a man adrift at sea.

"The helicopters will be coming from the north," the search pilot said.

"Tell us when they're ten minutes away," Bowersox said, nodding, mostly to himself. "We'll make sure they see us."

And with that, Expedition Six was all but over. All that was left was for Bowersox, Budarin, and Pettit to disband. But not just yet.

Bowersox told his friends that their jobs were finished and they could be happy again. He put down the radio, fell back into the dirt, and closed his eyes. Budarin dropped the shotgun, and Pettit stopped mentally turning their parachute into a tent, and they joined Bowersox in his meditation. Together, they put their backs flat against the grass once more, watching again the white birds that continued to circle overhead. They smiled at the thought of holding their wives and children, of having a hot shower, of slipping into some clean clothes, of putting their heads on pillows and pulling blankets up under their chins. They looked forward to waking from deep, perfect sleeps to their old gravity-bound lives, to becoming the husbands and fathers they once were and soon would be again.

But they also savored the silence. They savored those last honest moments of being alone. And by the time the beat of the helicopter's blades thumped over the horizon and the weight of the world had found their chests for the second time in the same afternoon, Bowersox, Budarin, and Pettit had each let his mind loose, floating, untethered, 250 miles up into the nothingness.

Each of them suddenly felt too far from home.

. . .

Bill Gerstenmaier's helicopter had returned to Astana for fuel. He had spent the past few hours hanging out of one of its windows, fruitlessly searching for Expedition Six. Now grounded and waiting for the gas tanks to fill, he had never been more desperate for a goldfish moment. Finally, the helicopter's radio crackled.

Mike Foale—one of the Mir astronauts, flying in another of the helicopters as NASA 's representative—and American flight surgeon Mike Duncan were pleased to announce that they could see Bowersox, Budarin, and Pettit, these three pale men who had turned into salvation. They were almost translucent in the sun, but they looked well and relaxed, as though they were in a shared trance. Foale and Duncan had waved down to them, and it took a moment for them to wave back, almost reluctantly. It looked as though they were waving goodbye rather than hello.

After their helicopter touched down on the grass, its wash blowing dust up and into the wind, Foale and Duncan joined the posse of soldiers, technicians, and medical personnel who had made the short run across to the astronauts. After some good-natured ribbing— What took you so long?—there were hugs and broad smiles. For Expedition Six, even the gentlest embrace felt like a vise, but it was a good kind of squeeze. It was something to see other faces.

Back in Astana, flush with joy and relief, Gerstenmaier remembered the anxious crowds waiting for his word in Moscow. He fired up his satellite phone and exalted in reporting a happy reunion: they have been found, and they have been seen, and they are very much alive.

His trembling voice was relayed through Houston and flung back toward two rooms in TsUP. In that magical instant, all of the time and all of the distance had finally disappeared, and the tension and anxiety and pain that had built up over five and a half terrible hours and however many miles was broken.

· · ·

The American delegation exploded in relief, Gerstenmaier's phone call having lit some kind of fuse. There were tears and hugs and handshakes and backslaps. Micki collapsed into Annie's arms. O'Keefe let loose with a victory howl. Readdy smiled and shook his head. Pastorek exhaled and put down his pen.

The Russians had also heard the news—as well as the roaring of the Americans—and emerged, cheering, from their own storm cellars. Having been divided by language and walls and now being united by celebration, the two camps met somewhere in the middle, in a massive conference room, the only room at TsUP big enough to contain the elation. There were the universal expressions of ecstasy: smiles and laughter, more handshakes and hugs. And then, from nowhere, several bottles of clear, cold vodka were broken out.

Almost immediately, a series of Russian officials began tearing into long, elaborate speeches, each of which was punctuated with a toast, a cheer, and another shot of vodka. The Americans caught on to the routine fast enough, and soon the speeches grew shorter and the glasses more full. In no time at all, everybody in that loud, bursting room was some level of drunk.

There might have been something base-seeming about that, in capping something as spectacular as the return of three men from space with a mass downing of alcohol. But those bottles of vodka were more than simple pollution.

For the men of Expedition Six, there had been nobler outlets to vent the anxiety that had followed them down to the flats. They could find ecstasy in the green of the grass and inhale gallons of crisp, clean air and watch white birds. In short, they could find in themselves a more poetic finish because their journey had come to such a beautiful end. They had known all along that they were safe.

But for the people who had been waiting so anxiously for their arrival, the day's final exclamation point had seemed less sweet. All they had been able to do was wait and worry, enduring hours spent tugged through a knothole. And while their own journeys had been just as dramatic as Expedition Six's, because they were static, because there had been no thump that signaled that they had made it back alive, they had felt robbed somehow. They were celebrating an abstraction. For them, each shot of vodka was the only reminder that all of their waiting and worrying was over. Each overturned glass and unmopped spill made out like gravity did for Expedition Six: each gave their patrons a heavy kind of comfort, the collective understanding that everything was going to be okay.

The toasting session's buzz continued long after the Americans had left to board a couple of minivans destined for Star City, where Bowersox, Budarin, and Pettit would arrive in four or five hours. For the people smiling out from inside the convoy, that drive through the suburbs of Moscow and into the trees went both fast and slow. As much as they wanted to see the objects of their affection with their own eyes, part of them also wanted the dreamlike anticipation to last, knowing that what had been so far away for so long was now so close.

. . .

When the space shuttle hits the runway, that's it. It's all over. There is the air, and there is the ground, and for the past twenty-five years, American astronauts have never had the chance to enjoy the space in between. But for the men who used to be Expedition Six, they basked in this more graceful transition. As if to remind them of what they once were and where they had once been, Bowersox, Budarin, and Pettit were soon back in the air. It's one of the joys of returning by capsule—as it was with *Apollo*, so it is with *Soyuz*. Always, touchdown is followed by another liftoff, this time to a military installation in gloomy Astana, and then in the Aeroflot jet that would take them to Star City. They were no longer spacemen, and that truth struck them hard, but they remained fliers. Through their windows, they could still look down on the Kazakh steppes, the way they had looked down on them through their windows back on station, thinking they were beautiful then and thinking they were beautiful now.

. . .

A few hundred people—although only a few that mattered most—were waiting for Bowersox, Budarin, and Pettit when their plane finally touched down. The reception committee had filled the hours with more vodka and a little sleep, and Micki had also given herself time for a good long cry on her cottage step. It was the same cottage in which she had been singing with friends when the phone had rung so many months ago, calling her husband away. After she had cleaned herself up, she and Annie were packed onto a bus and given red roses and told to watch for their men through the windows.

The plane made a slow turn and taxied toward the crowd. There were cameramen and dignitaries and soldiers, the bunch of them getting damp in the mist that continued to fall out of a thin gray sky.

Whether it was for nostalgia's sake or because of budget constraints, the Russians wheeled up one of those old-fashioned staircases. Perhaps it was because the setup made for such terrific pictures. Reporters and photographers stood poised for what seemed like forever, waiting for the plane's engines to stop and the door to open. Finally, it was cracked like a hatch.

So, too, was the door of the bus, and Micki and Annie stepped out into the rain and the bright white of camera flashes. They strained their necks above the crowd to try to catch sight of their husbands, although Micki had already been told that she would likely need to come to Don rather than the other way around.

As commander, Bowersox had the honor of being the first out, smiling and raising his fists in the air. Boosted by a last-second charge of adrenaline, he ran down the stairs and launched himself into the arms of the surging crowd. Sean O'Keefe had grabbed hold of Annie Bowersox's hand, but now they nearly lost each other in the crush. It was like a rock concert after the lead singer takes a dive from the stage.

Budarin was next. He came down a little more slowly, a little more proudly, but waving and smiling, too.

After the commotion had died down, after the bottom of the stairs had been cleared, Micki made her way up. She ducked into the plane and saw Don, curled up in a corner under a blanket. He looked at her as though he had wanted to see no one else, and she put her hand to her mouth, not quite believing.

They had time only for a kiss. Pettit, having refused the offer of a stretcher or a chair, was helped out of the plane and down the stairs by two men. His jelly legs belied his good feeling.

The three men were wrestled through the throng and guided back to the bus, still idling in the near distance. O'Keefe pushed through the crowd, towing Annie Bowersox in his wake, and the two of them knocked together on the glass door. Bill Readdy had taken Micki Pettit by the arm, and they, too, made it through the celebration and into the relative sanctuary of the bus. Through the rain-streaked windows, they watched the crowd continue to cheer and applaud, but now, inside, there was only the hush of relief broken by short outbursts of joy. Best of all, there were long hugs between husbands and wives who had thought in their weaker moments that another touch might never come. They kissed and put their heads on each other's shoulders, and they swayed with the lurching of the bus.

It was the sort of poignant scene that films usually close on, fading out on a New York City sidewalk or at the bow of a ship at sunset or in the arrivals terminal of a busy airport. Music swells up, and the camera pulls back, revealing two people locked in a tight embrace, perfectly still amidst the chaos that swirls around them, together, alone against the universe. That's exactly what it looked like when that bus threaded its way through the crowd and made for open range. The only things missing were the names of the key grips, best boys, and set designers projected against the wet windows.

But that's not how these stories end in real life. The finish is never quite so neat. There are always footnotes.

. . .

Over the rumble of the diesel engine, Sean O'Keefe pulled out his cell phone and called up Washington, D.C., where Vice President Cheney had been waiting for word of his former employee's success. (He even put Bowersox on the line to say hello.)

Nikolai Budarin looked forward to seeing his own wife, Marina, who was also waiting, but with an infusion of love, cigarettes, and vodka.

Waiting for them, too, was a gang of scrubbed-down flight surgeons cloistered inside the Prophy, the Russian postflight checkout building, a kind of one-stop hotel and hospital for returning spacemen. The doctors were poised to get their gloved hands on three exhausted men who had made the transition from astronauts to experiments and now, finally, to fresh specimens, ripe for dissection. The bus stopped at their open front doors. With the hiss of brakes, reunion was interrupted.

Bowersox, Budarin, and Pettit were escorted inside. They were treated gently by the Russians, who have learned over the years how tender their subjects can be, but the three men were put under the microscope all the same. Like every astronaut and every cosmonaut, from the first to the last, they were seen as something alien and wonderful, these ordinary assemblies of skin and tissue that had been turned into artifacts by virtue of the places they had been. Spending nearly six months in space had made them worthy of a fine-eyed examination, and now they were looked upon if not as heroes, then at least as monuments.

However subtle the adulation was, it made Bowersox, Budarin, and Pettit feel vaguely uncomfortable. They were still adjusting to gravity's weight and the taste of earth's air. The whiff of something as ordinary as muscle rub hit them like smelling salts. Ringing phones sounded like fire alarms. A change in temperature of just a few degrees left them either shivering or burning up. And on top of all of that, now here they were, after having spent so much time alone, stuck in the center of a clutch of curious physicians, measuring their fat stores and shining lights into their eyes.

The most painful invasion was saved for last, nearly thirty-six hours after Expedition Six had blinked awake for their final morning in space. The surgeons took liberal biopsies of their calf muscles, leaving three divots in each of the men. The open sores were reminders for Bowersox, Budarin, and Pettit that it would take months and perhaps even years for them to earn back the calluses that they had lost; their skin was still fresh and pink. In the meantime, they tried to separate themselves from the noise and commotion—they tried to look inside themselves for places to hide, to insulate themselves from their new surroundings. They were only partly successful. They felt as though they had been set upon, as though these bizarre masked crusaders had broken through their best defenses, intent on shaking them out of their comas.

And worse, they knew, too, that this was just the first of it.

They began steeling themselves for the chaos to come. By the time they emerged from the examination rooms, they had already grown harder. Already, they had begun patching the workaday glaze that had taken them so very long to shed.

They were placed in quarantine, as much for their benefit as for anybody else's. First, though, Budarin had time to sneak out to see his wife and collect his smokes. O'Keefe and Annie sat down with Bowersox for a short, filmed chat about their extraordinary return to earth. ("Nikolai was like a cheerleader up there!" Bowersox said.) And Micki was given a few minutes to visit with Don, who had been laid out in bed. He was drifting in and out of sleep.

She made sure he was tucked in, his blanket pulled up, and she planted another gentle kiss on his forehead. She wished him good night. "Get some rest," she whispered. She said she would try to see him in the morning if they would let her, and with that, she stepped out into the hallway.

Looking back through her husband's door, she felt as though she had said good night to a man she had never seen before but also to a man who had never gone away.

. . .

While Bowersox, Budarin, and Pettit remained locked down, their friends and family drowned themselves in more vodka. There was a sweet spread waiting outside of the Star City cottages that had been reserved for them. Chairs and tables were scattered on the grass. There were baskets of bread and bowls of salad. Meat was crackling over coals and under skies that had just started to clear. Everybody filled their plates and grabbed a drink and sat down heavily into their seats or stood together in clusters, basking in one another's glow. It was like a long-overdue family reunion, except that three of the most honored guests were absent, sixteen dawns and dusks a shift having given way to one. The sun set. And with it, the gathering outside the cottages broke up. Bleary and wrung out, the partygoers drifted off into the night, one by one, two by two.

Sean O'Keefe, Bill Readdy, Paul Pastorek, and a few other NASA officials weren't quite ready for sleep, however. They were still coming down from a long day, swings of jet lag and adrenaline that had made like a speedball. They voted to retreat to the basement of one of the cottages, to a small room that had been dubbed Shep's Bar.

It was named in honor of Bill Shepherd, U.S. Navy SEAL and commander of Expedition One. To quell his homesickness during his years of training—and to quench his thirst for something other than another shot of Russian vodka—he had set up this makeshift hole-in-the-wall, a kind of speakeasy for the young astronaut set. It wasn't much, a few chairs and tables in the dark and dank, but it harbored perhaps the most valuable piece of hardware in all of Star City: this great, shining, magnificent machine that turned ice and alcohol into frozen daiquiris.

It had been rustled up by a man named John McBrine, one of NASA's longest-serving staff members in Russia. He had found the monster for sale on the Internet, but after he had unpacked it,

McBrine was crushed to find that the drinks it spewed out were warm. Fortunately, an American reserve astronaut named Don Pettit, ignorant to the adventures that awaited him, had a knack for fixing just about anything that had been broken. He also had some time on his hands. The way Pettit would later repair a certain Microgravity Glovebox on the International Space Station, he repaired McBrine's daiquiri machine.

Now that machine was fired up, shaking on the counter, rumbling like one of that afternoon's helicopters, and out of it poured an endless stream of boozy froth. Loud and happy men gathered it up and pushed it back, in between still more hugging and hollering, blowing off the last traces of steam that had been generated over the past hours, days, weeks, and months. For most of the men packed inside that bar on that night, it was the first grand good time they had enjoyed since *Columbia* had cast them in shadows. The weight of the world had been lifted off their shoulders just as Bowersox, Budarin, and Pettit had assumed it. For the first time since February, they felt free, and they would have liked the feeling and the night to last forever.

Eventually, though, morning came, and with it short naps and trips to the airport, where flights waited to carry the men on their long rides home. On one of them, O'Keefe and Pastorek sat side by side, grinning through hangovers and fatigue. There were no more notes to take, no more decisions to make, and best of all, there was no more sleeplessness. Before their plane was even wheels up, they had closed their eyes and drifted off into a sleep so deep, they needed only to give a little kick, and they, too, would know how it felt to fly over mountaintops and look down on skyscrapers.

. . .

The rest of them—Bowersox and Annie, Pettit and Micki and their kids—woke up in separate beds, in separate rooms, to three more weeks in Russia. They enjoyed the occasional visit together, but each was short, twenty minutes one afternoon, a ten-spot the following morning. The astronauts spent the rest of their days being examined by doctors and interviewed by experts about their experience, most energetically about their *Soyuz* flight. The engineers charged with figuring out why it had gone ballistic believed, openly, that the crew was at fault. Again and again, the three men denied any wrongdoing. They had been crew members on a ship and in a universe each with a mind of its own. They had been at the mercy of something outside of themselves.

And now they were again. They were back on earth, back to all of its pressures, back to all of its demands on their time and their bodies. They sometimes felt like wrestlers who had been pinned, helpless, to the mat, their fates no longer their own.

From the moment they had pulled up to the gates at Star City, their days had stopped unfolding exactly as they wanted them to. For the first time in nearly six months, they had to alter their routine to make room for dozens, hundreds, and even thousands of others in it.

Soon they would be caught in traffic and the rain, bumped into on the sidewalk, jostled on the subway, tied to a desk for hours each day. They would catch colds. They would have appointments to keep, and the grass would need cutting.

They would be rushed. They would be late.

And in the quiet in between, they would wonder at what they had done, and they would wonder at what's next.

EPILOGUE

Mars.

In January 2004, a little less than nine months after Expedition Six's dramatic fall to earth, President Bush outlined a brave new vision for NASA. (He was introduced from station by Expedition Eight's Mike Foale, the *Soyuz* taxi missions having worked perfectly since Ken Bowersox, Nikolai Budarin, and Don Pettit had their brush with death by gravity.) To surprisingly little fanfare, the president called for the completion of a scaled-down International Space Station by 2010, a return to the moon as early as 2015, and a manned mission to the red planet sometime after 2020.

While the rest of the country tried to a stifle a collective yawn, a large pocket of Houston seemed grateful for the spark. Still struggling to fix the external tank's insulating foam—the same foam that had doomed *Columbia*—the agency's technicians and engineers were suddenly presented with a grander challenge, exactly the sort of rock-solid objective that they had been itchy for.

The announcement awakened something within the men of Expedition Six, too. In the months that followed their bumpy ride home, they were cleared of any wrongdoing in their capsule's malfunction. Budarin had been adamant all along that the crew was not at fault; Bowersox was less certain. ("One thing I've learned in flying airplanes over the years is never say for sure that you didn't make a mistake," he said. "It's always best to be humble.") After the investigation was complete, it was found that a run-of-the-mill software glitch had been the root of so much drama.

It was a shock to learn that their lives had been at the mercy of some tired computer programmer's typo. But Bowersox, Budarin, and Pettit also saw something fortuitous in their misadventure, especially given the president's proclamation—the perfect finishing note for an expedition that had been built on the science of accident.

As though by design, their extended mission and eventful return mirrored the long trip to Mars as closely as any journey into space ever had. The gutsy astronauts chosen to make that incredible voyage would need to lift off in a rocket, live in weightlessness for six lonely months, burn down through the planet's atmosphere in a capsule, dig into its rocky surface, and finally tap their innermost strength, cracking open their hatch and setting up camp.

Mission Control would never have taken the risk of purposely leaving Expedition Six to fend for themselves on the Kazakh steppes. But that inadvertent cold shoulder had proved that perhaps the hardest part of a Martian landing was, in fact, possible. Even if little green men attacked the newly landed crew, Budarin had demonstrated that the astronauts would have it in them to pump their double-barreled shotguns and blast buckshot into the shadows. Bowersox, Budarin, and Pettit had shown that, if nothing else, our astronauts possess the necessary fight.

They were rewarded for their groundbreaking troubles. Not long after Bowersox had settled back into Houston's routine, he was named director of NASA's Flight Crew Operations Directorate. His new job, among the most powerful positions in the astronaut office, includes selecting which men and women will occupy which seats on which flights.

Budarin was promoted to the position of flight director. In his new role, he jets between Houston and Moscow, helping to manage the American and Russian spaceflight programs. He is happy for spending so much time stateside. In Russian culture, friendship is especially meaningful, and he delights in keeping Expedition Six nearly as close on the ground as they were in space. Whenever Bowersox, Budarin, and Pettit sit together around a restaurant table, laughing and remembering, they never try to shake the sensation that the rest of the room might fade into black and they will be left feeling like three men sitting in a bucket, alone against the universe once again.

Still, the feeling needs constant nurturing: given their new roles, it is unlikely that Bowersox and Budarin will ever fly again. Only Pettit remains in the active astronaut pool, filling his time served on the ground by working on the shuttle's foam problem and helping design the Crew Exploration Vehicle, or CEV, the ship that is expected to replace the space shuttle when the fleet is retired in 2010.

Asked by his managers whether he wanted to be assigned to one of the few remaining shuttle flights, Pettit declined. Instead, he asked to be considered for another long-duration mission to station. Micki wasn't entirely thrilled with his request, but his managers smiled, nodded their heads, and put his name down on the list.

That list remains long. The shuttle didn't return to space until *Discovery* launched in July 2005, two and a half years after *Columbia*'s final flight. (It was a good thing that Sean O'Keefe and company hadn't settled for the Avdeyev Option; Bowersox or Pettit might have come back with gills or to empty houses.) A new camera revealed that a large piece of insulating foam had still fallen from the external tank, and the fleet was grounded again. With only two-man crews occupying the International Space Station for six months at a stretch, and with the three remaining shuttles empty and locked in their hangars, NASA's astronaut office started to feel less like an airport lounge and more like a prison without bars.

The boxed-in feeling has been especially strong for Pettit. Bowersox and Budarin have been able to reconcile themselves to their likely permanent grounding. The American has rocketed into space five times, and the Russian has clocked three long-duration missions. They would have liked more, but each man knows that he has been given his time.

But Pettit has been up there only once, and he feels as though there is still so much for him to do. Every day he spends trapped on earth, some part of him feels denied.

Now he understands what the other flown astronauts in his class went through upon their return. They had always seemed unsettled, lost in conversation, distracted. They were always caught dreaming. They had forgotten their fear, and they had forgotten their terrible solitude. But what had been hardwired into them, seemingly for good, was their longing. In the days after they had returned, they had begun their scheming, trying to figure out how to make it back to the place where they now believed they most belonged. And they had come to know what only their fellow astronauts—from Armstrong, Aldrin, and Collins on—could understand: that their gold astronaut wings sometimes felt like a scarlet letter.

They were different from the rest of us now, and different, too, from the men they once were, and not just because so many of them still swallowed their toothpaste.

Before, Pettit would stare up at the sky and feel as though the stars were close enough for him to touch. Now he no longer finds comfort in that easy lie. Even when he checks his watch, sneaks through his front door late at night, sets up his telescope on his lawn, and follows the space station on its long journey through the universe, he can feel cut off from his home as if by a wall, a wall as thin as a single sheet of glass.

ACKNOWLEDGMENTS

First and foremost, I'd like to acknowledge Ken Bowersox, Don Pettit, and Nikolai Budarin, the three brave men whose stories I've tried my best to tell in this book. I would especially like to thank Don and his terrific wife, Micki, who were particularly generous with their time.

Their cooperation began when I first wrote their story as a feature for *Esquire* magazine, where I'll be delighted to work for as long as they'll have me. I was brought on board there by Andy Ward, who has since moved on to *GQ*; I will forgive him for that, because none of this would have happened if he hadn't let some dumb kid bearing a box of Krispy Kremes into his office one summer afternoon. In a cosmic way, Andy is the start of all of this, and I am hugely indebted to him.

I also owe a giant-size debt to my eternally patient editor at *Esquire*, Peter Griffin, who has, in his quiet way, made me a much better writer in the time that I have worked for him. He has let me see through his eyes what makes for a good story and how best to tell it, including some incredibly valuable advice when I wrote about Expedition Six for the magazine. Peter also supported my writing *Too Far from Home* throughout its long, tortured birth. Even if it was just a quick phone call to ask how things were coming along, he spurred me.

Of course, I must give a particularly robust thanks to David Granger, the editor in chief of *Esquire* and my boss of bosses. Not only is he generous enough to continue employing me (at least as of this writing) but he wrote one of the kindest e-mails a writer could imagine receiving after he read my Expedition Six story. I've kept it and read it often in the nearly two years since.

As well, David did me the service of passing my first draft along to Bill Thomas, my Great Benefactor—publisher and editor, more officially—at Doubleday in New York. I'm not really sure what happened next. All I know is, I was covering the Masters in Augusta, Georgia, when I received an e-mail from my agent, David Black: "Your life just changed," he wrote. "Call me." I did, and he gave me the unexpected news that he had sold a book-length version of the story to Bill. Golf never saw so much screaming. These three men—Granger, Thomas, and Black—conspired to give me a thrill that I will never forget. They each share a permanent stamp in my great memories passport.

Shortly thereafter, the dreamy fantasy stopped and the work began. Although NASA declined to help me out—for reasons I've never been able to fathom—many others did, and for that, I am grateful. (Luckily, several of the people I needed most have left Houston behind.) Sean O'Keefe, now the president of Louisiana State University, was especially frank in his recollections. His friend and lawyer, Paul Pastorek, elected not to charge me by the hour, and I appreciate that. Bill Readdy, now having started up his own consulting firm, Discovery Partners International, also took breaks from his busy schedule to share his memories of bad days and good.

Among the probably dozens of others I've leaned on, I would also like to thank Christine Pride, Karla Eoff, Todd Doughty, and everyone else at Doubleday; Doris Lance at the Naval Air Warfare Center in China Lake, California, and former pilot Dick Wright; Deborah Goode at the United States Naval Academy; Dr. Tom Peterson and his assistant, Gerri Sullivan, at the University of Arizona in Tucson; John Haire at Edwards Air Force Base; Konstantin Tyurkin of RIA Novosti in Moscow; Sergei Gruzdev, my smiling and able translator; Neil Woodward, American astronaut; and the staff at the Ottawa Public Library, including my wife, Lee.

The librarians helped me find the many books that sat in piles on my desk, tall as pillars. To ignite things, I started out by reading two of the best: Tom Wolfe's *The Right Stuff*, and Norman Mailer's

Of a Fire on the Moon. Really, reading them just depressed the hell out of me, but at least they got me going. Of the hundreds of other books, magazine pieces, and newspaper articles I have read and sometimes stolen from (but always with a suitable measure of shame), I would like to make particular mention of the most illuminating: *Comm Check . . . The Final Flight of Shuttle Columbia*, by Michael Cabbage and William Harwood; *Columbia—Final Voyage*, by Philip Chien; *Red Star in Orbit*, by James E. Oberg; *Off the Planet: Surviving Five Perilous Months Aboard the Space Station Mir*, by Jerry Linenger; and *Dragonfly: NASA and the Crisis Aboard Mir*, by Bryan Burroughs (although I suspect that Bryan's book in part made NASA less trusting of my own intentions, so I feel like, if anything, he owes me).

I also wish to give a loud and long shout-out to the awesome resource that is Space.com, particularly the first-rate reporting of Jim Banke. His work helped me reconstruct the days before *Endeavour*'s belated launch with what would have been an otherwise impossible level of detail.

I'd like to thank the Weakerthans, the finest band ever to come out of Winnipeg, Manitoba, and perhaps the world. Their albums served as much of the inspiration for my writing this book, and I also lifted a phrase for the title of Chapter 6 from "Left and Leaving," which is one of their great songs.

No one has cared more about this book than my family, who have lived and died with it right along with me. My loving parents, John and Marilyn, provided a much-needed mix of encouragement, advice, and copy-editing expertise. My brother, Steffan, was an early supporter and reader, as were my parents-in-law, Jim and Alice Higginson. (My best friend, Phil Russell, also slogged through the first draft of the book; never have so many remarks in the margins included the salutation "dude.") And you can never underestimate the amount of a book's weight that is shouldered by a spouse. Perfect Lee helped float me through some trying times, and she never once made me feel bad for failing to hear her calling when I was lost in space.

Not only that, but I wrote the final few pages of *Too Far from*

Home in the maternity ward at Ottawa's General Hospital, sitting next to Lee, who was confined to a white-sheeted bed, heavily pregnant. She was in the hospital for three weeks before our first son, Charley, was born along with the last words on these pages. It was an exhausting, exhilarating, gut-crazy time. It was, I like to imagine, exactly like riding a rocket.

ABOUT THE AUTHOR

CHRIS JONES was a sportswriter at the *National Post*, where he won an award as Canada's outstanding young journalist. He joined *Esquire* as a contributing editor and sports columnist, and became a writer at large when he won the 2005 National Magazine Award for Feature Writing for the story that became the basis for this book. His work has also appeared in *The Best American Magazine Writing* and *The Best American Sports Writing* anthologies. He lives in Ottawa, Canada.